CAMBRIDGE LIBRARY COLLECTION

Books of enduring scholarly value

Astronomy

From ancient times, humans have tried to understand the workings of the world around them. The roots of modern physical science go back to the very earliest mechanical devices such as levers and rollers, the mixing of paints and dyes, and the importance of the heavenly bodies in early religious observance and navigation. The physical sciences as we know them today began to emerge as independent academic subjects during the early modern period, in the work of Newton and other 'natural philosophers', and numerous sub-disciplines developed during the centuries that followed. This part of the Cambridge Library Collection is devoted to landmark publications in this area which will be of interest to historians of science concerned with individual scientists, particular discoveries, and advances in scientific method, or with the establishment and development of scientific institutions around the world.

Observations of Comets from B.C. 611 to A.D. 1640

Long before their western counterparts, Chinese astronomers developed remarkably accurate methods for making their measurements, recording detailed observations since ancient times. Of particular interest to John Williams (1797–1874), assistant secretary to the Royal Astronomical Society, were Chinese observations of comets. Noting that previous translations of these records had been incomplete, Williams sought to produce a fuller catalogue. The present work, published in 1871, presents Chinese comet observations between 611 BCE and 1640 CE, using the encyclopaedia of Ma Duanlin and the great historical *Shiji* as major references. Williams provides useful context in his introductory remarks, mentioning the tests by which the accuracy of the Chinese records can be verified. He also includes chronological tables and a Chinese celestial atlas, enabling comparison between the Chinese and Western systems for dates and stars respectively.

Cambridge University Press has long been a pioneer in the reissuing of out-of-print titles from its own backlist, producing digital reprints of books that are still sought after by scholars and students but could not be reprinted economically using traditional technology. The Cambridge Library Collection extends this activity to a wider range of books which are still of importance to researchers and professionals, either for the source material they contain, or as landmarks in the history of their academic discipline.

Drawing from the world-renowned collections in the Cambridge University Library and other partner libraries, and guided by the advice of experts in each subject area, Cambridge University Press is using state-of-the-art scanning machines in its own Printing House to capture the content of each book selected for inclusion. The files are processed to give a consistently clear, crisp image, and the books finished to the high quality standard for which the Press is recognised around the world. The latest print-on-demand technology ensures that the books will remain available indefinitely, and that orders for single or multiple copies can quickly be supplied.

The Cambridge Library Collection brings back to life books of enduring scholarly value (including out-of-copyright works originally issued by other publishers) across a wide range of disciplines in the humanities and social sciences and in science and technology.

Observations of Comets
from B.C. 611 to A.D. 1640

Extracted from the Chinese Annals,
Translated with Introductory Remarks

JOHN WILLIAMS

CAMBRIDGE
UNIVERSITY PRESS

CAMBRIDGE
UNIVERSITY PRESS

University Printing House, Cambridge, CB2 8BS, United Kingdom

Cambridge University Press is part of the University of Cambridge.
It furthers the University's mission by disseminating knowledge in the pursuit of
education, learning and research at the highest international levels of excellence.

www.cambridge.org
Information on this title: www.cambridge.org/9781108078115

This edition first published 1871
This digitally printed version 2014

ISBN 978-1-108-07811-5 Paperback

OBSERVATIONS OF COMETS,

From B.C. 611 to A.D. 1640.

OBSERVATIONS OF COMETS,

FROM B.C. 611 TO A.D. 1640.

EXTRACTED FROM THE CHINESE ANNALS.

TRANSLATED, WITH INTRODUCTORY REMARKS,

AND

An Appendix,

COMPRISING THE

TABLES NECESSARY FOR REDUCING CHINESE TIME TO EUROPEAN RECKONING;
AND A CHINESE CELESTIAL ATLAS.

BY

JOHN WILLIAMS, F.S.A.

ASSISTANT SECRETARY OF THE ROYAL ASTRONOMICAL SOCIETY,
ETC. ETC.

LONDON:
Printed for the Author
BY
STRANGEWAYS AND WALDEN, CASTLE STREET, LEICESTER SQUARE.
1871.

TO

WILLIAM LASSELL, ESQ. F.R.S.

PRESIDENT,

AND TO THE FELLOWS OF THE ROYAL ASTRONOMICAL SOCIETY,

THIS WORK

IS RESPECTFULLY INSCRIBED.

ERRATA and CORRIGENDA.

INTRODUCTORY REMARKS.

Page x, line 25, *for* There *read* These.

xi,   „   31, *for* central one *read* central ones.

„   „   32, *after* 'Monthly Notices' *insert* of the Royal Astronomical Society.

xiii, lines 20, 41, 43, *for* B.C. 613 *read* B.C. 611.

xix, line 32, *for* Idler *read* Ideler.

xx,   „   22, *for* after *read* above.

xxviii,   „   7, *for* and *α* Andromedæ *read* and stars near.

xxix, last line, *for* computation *read* compilation.

xxx, line 3, *for* Zeitrichnung *read* Zeitrechnung.

OBSERVATIONS.

6, lines 21 and 24, *after* B.C. 110 *add* June.

33, line 5 from bottom, *for* January 6 *read* January 16.

66,   „   28, *for* Seu *read* Seuh.

67,   „   30, *for* 3rd year *read* 5th year.

TABLES.

105, line 1, *for* Tung Hang *read* Tung Han.

112, lines 5 and 6, *for* 袖 *read* 祐·

In the twenty-four divisions of the Chinese year in Table H, the asterisks referred to in p. xxiii, line 1, of the Introductory Remarks were accidentally omitted. They should be placed before every second line commencing with Ta Han, Jan. 21, and ending with Tung Che, Dec. 22.

PREFACE.

A FEW years since, when Sig. Schiaparelli announced his discovery of the probable connexion between the orbits of certain Comets and those of the periodical Meteors, the well-known Astronomer, Mr. J. R. Hind, took up the subject, and in the course of his investigation found, among the comets recorded in M. E. Biot's 'Catalogue des Comètes observées en Chine depuis l'an 1230 jusqu'à l'an 1640 de notre ère,' &c., which forms a supplement to the 'Connaissance des Temps' for 1846, one, whose orbit appeared to answer the required conditions. It is that of October 25, 1366, and is No. 295 of the succeeding Catalogue. But the path of that comet, as given by Biot, was apparently so extraordinary, that he was induced to conclude that some error had occurred, either in the original text or in the translation. Thus the comet is described as having appeared near one of the stars in Ursa Major, whence it passed in a south-easterly direction through several of the stellar divisions, until, on October 29, it was in a certain place in Aquarius; but on October 30 it was again in Ursa Major, in the same place as at first : whence it once more pursued a similar south-easterly course to the same place in Aquarius and disappeared.

Mr. Hind consequently applied to me, to know whether this discrepancy could be accounted for by reference to the original text in the Supplement to the 'Encyclopædia' of Ma Twan Lin. Upon carefully examining this, I found that there were really *two* distinct accounts of the course of this comet ; the one giving its path through the stellar divisions, and the other describing the asterisms near or through which it passed in that course ; and, reading the account according to this version, the whole became perfectly consistent, and Mr. Hind was enabled to bring his investigation to a satisfactory close.

This led to an examination of Biot's Catalogue, and I quickly found, that although very accurate in its details, it was by no means so complete as

a

could be wished ; many comets being recorded in the ' Encyclopædia ' of Ma Twan Lin, and in the great historical work called the ' She Ke,' that are not noticed by him. It therefore appeared to me, that a Catalogue comprising the whole of the observations of the comets contained in the two Chinese works just mentioned, translated from the original, and arranged chronologically, with an explanation of all the particulars connected with them, might be of some service to astronomers, particularly to those engaged in cometary researches.

Such was the origin of the present work ; and as during its compilation many other interesting particulars relating to Chinese astronomy presented themselves, I have been induced to incorporate the principal of them with the Introductory Remarks, which immediately follow.

It was likewise absolutely necessary for the finding the dates, &c. of the various observations, that certain Tables should be constructed, by which Chinese time could be reduced to our reckoning. These will be found in the Appendix, and consist of a complete set of Chronological Tables, giving the succession of the Dynasties and Emperors from the earliest period to the present time ; and of other Tables for finding the Months or Moons, and Days. Instructions for using these Tables are given in the Introductory Remarks, and they will be found of great service, not only to astronomers but also to persons engaged in historical or numismatic investigations, as they will find in them all they require to ascertain the various dates, as far as regards China proper. I have also included in the Appendix a complete Chinese Celestial Atlas, from an original work, so that the names and relative positions of the asterisms and stars can be readily found.

I may also express my conviction that this Catalogue will be found as complete as any that has hitherto appeared, if not more so. It has likewise the advantage of being a work *per se*, and, consequently, will appear in the library or in catalogues as a separate work, and not as forming a portion of any other publication.

INTRODUCTORY REMARKS.

THE progress of Astronomy among the Chinese is a subject of the highest interest, whether it be considered as recording observations of the heavenly bodies made by one of the most ancient and primitive races of mankind, which appears in extremely remote times to have advanced to a high degree of civilisation; peculiar, however, to itself; and which has preserved the manners and customs established by its early rulers, more than two thousand years before the Christian era, in a great measure unaltered to the present day. Or whether the fact that at a period long anterior to the commencement of civilisation among the Western nations, and when (with the exception, perhaps, of the Egyptians and Assyrians) almost universal barbarism prevailed among them, Astronomy had been carried to a great degree of perfection by the Chinese, as manifested by their still existing records, whose authenticity is not only strongly asserted by that people, but is acknowledged by some of the most eminent European scholars of the present day.

In their later records, in addition to a vast amount of valuable historical and other information, we find chapters devoted exclusively to their astronomy: in which are detailed their arrangement and classification of the stars; observations of the sun, moon, and five planets; notices of eclipses, falling stars, and other extraordinary phenomena: among which those of comets, which are extremely numerous, and which frequently are minute in the description of the times and places of their appearance, and of the paths they pursued in the heavens, are perhaps the most interesting to modern astronomers.

As far as my experience goes, it is not easy to find a complete record of these observations in any European language; and if such does exist, it is contained in publications not readily accessible to the general reader. Pingré, in his 'Cométographie,' quotes lists of Chinese comets by Mailla and Gaubil. Mailla's list was taken from the Chinese historical work called 'Tung Keen Kang Muh,' which he translated, and of which I possess a copy, which is occasionally referred to in the following work. That by Gaubil is said by Pingré to have been in MS., and to have been preserved in the library of the Depôt de la Marine. I have been unable to ascertain whether this MS. is still in existence, or whether, as is highly probable, it was lost in the early days of the French Revolution. Under any circumstances, it does not appear at present to be easily accessible to the general inquirer. The important lists in the 'She Ke' and in the 'Encyclopædia' of Ma Twan Lin do not appear to have been known to Pingré. The Catalogue by M. E. Biot, published in the 'Connaissance des Temps' for 1846, although very accurate, is by no means so complete as could be wished, many observations of comets being passed over without notice.

The remarks which follow are intended to supply certain desiderata, which appear to be essential to the full comprehension of the observations which form the subject of this work. They consist principally of,—

Firstly, A brief account of the early astronomy of the Chinese, extracted entirely from original sources, chiefly historical.

Secondly, An explanation of the means to be employed in reducing Chinese time to our reckoning, including all particulars relating to the dates of the Emperors who were reigning when the comets described appeared, such as the years of their reigns and epochs; and the dates of the Moons or Months, and Days referred to in the descriptions, and an explanation of so much of the Chinese calendar as may be needed for the understanding the mode of their reduction; together with the requisite Tables for that purpose.

Thirdly, An explanation of the astronomical portion of the observations, showing the mode of ascertaining the various asterisms and stars mentioned as occurring in the paths of the comets, as they are described in the Chinese lists and maps of stars, with other particulars relating to them necessary to be noticed.

Fourthly, An explanation of the plan followed in the translation and arrangement of these observations.

These, it is confidently hoped, will render the work intelligible to the general reader.

According to Chinese tradition, the introduction of astronomical observations is to be attributed to Shin Nung, the immediate successor to Fuh He, the founder of the empire. The reign of Shin Nung commenced about B.C. 3253.

It is also related that the Emperor Hwang Te, B.C. 2698–2598, was the monarch who introduced the system of reckoning their chronology by the cycle of 60 years, which he regulated by means of two series of characters; the one of 10, the other of 12 characters, called, from the first of each series, Kea Tsze. This system is in use to the present day, and will be fully explained in a subsequent page. The year of the introduction of the cycle is the starting-point of Chinese chronology; and I may observe that the present cycle is reckoned by the Chinese as the 76th, the first year of which was A.D. 1864. It therefore follows, that in 1863 the 75th cycle was completed: consequently $75 \times 60 = 4500$, the number of years then elapsed from the first year of the first cycle; and $4500 - 1863 = 2637$, the date B.C. of that year, which is said to have been the 60th year of Hwang Te. He is also considered as the inventor or discoverer of the lunar cycle of 19 years, by which the return of the conjunctions and oppositions of the sun and moon can be calculated, and the intercalary moons regulated. Hence it should appear that the lunar cycle of 19 years, introduced among the Greeks, many ages after, by Meton, was known to the Chinese more than 2000 years before that astronomer was born. These are mentioned merely as Chinese traditions, and not as resting on any other authority.

In the Chinese annals it is recorded, that in the reign of Chuen Kuh, the grandson of Hwang Te, in the spring of the year, on the first day of the first moon, a conjunction of the five planets occurred in the heavens, in Ying Shih. Ying Shih, or, as it is more usually denominated, Shih, is one of the 28 stellar divisions, determined by a, β, and other stars in Pegasus, extending north and south from Cygnus to Piscis Australis, and east and west 17 degrees, and comprising parts of our signs Capricornus and Aquarius. The Emperor Chuen Kuh is said to have reigned 78 years, from B.C. 2513–2436, and to have died in his 97th year; and from modern computations (I believe, by Bailly the French astronomer) it has been asserted that a conjunction of the five planets actually *did* take place about the time and within the limits indicated, *i. e.* on the 29th of February, 2449 B.C., being the 65th year of Chuen Kuh. Should this, on further investigation, prove correct, it will afford a strong presumption of the authenticity of the early Chinese annals, as there is no appearance of their astronomers having been at any time able to compute the places of the planets so far back; and the account is found in works published long before any intercourse with Europeans had taken place.

The next notice of Chinese astronomy appears in the 'Shoo King,' one of their five classical works, which is considered by the Chinese as the most ancient of their books. We have it as revised by Confucius, about the sixth century before our era. It was even then considered as of extremely remote antiquity, and from the peculiarities of the style of the early portions of that work there is but little doubt of such being the fact. Not only do the Chinese themselves assert its authenticity, but many of the best European scholars believe it to be genuine. Among these, Father Gaubil expresses no doubt of that fact; and M. J. B. Biot, in his account of Chinese astronomy, informs us that M. Stanislas Julien, without doubt the most accomplished Sinologist in Europe, has expressed the same opinion, which he derives from internal evidence, founded on its peculiar archaic style and construction. On examination, it appears to be rather a collection of historical documents of different ages than a continuous history, and may be considered as being quite as much to be relied upon as any of the histories of ancient nations that have descended to us.

The first section of the 'Shoo King' is called Yaou Teen, and records the actions of the Emperor Yaou. According to the Chinese authorities, this prince ascended the throne in the 41st year of the 5th cycle of 60 years. It has just been shown, that the reckoning by cycles commences with the year B.C. 2637. Four cycles of 60 years will be 240 years, and 41 years of another cycle will make 281; consequently 2637 less 281 will give 2356 for the first year of Yaou.

In the first section of the 'Shoo King,' just mentioned, in the paragraphs 3 to 8 inclusive, the instructions of Yaou to his astronomers, under the designations of He and Ho, are given. These names, He and Ho, are possibly not those of individuals but of two families, under whose superintendence the arrangement of the calendar for the year, and the making the necessary observations and the requisite computations, were placed, and whose office, in accordance with Oriental custom, was probably hereditary. These instructions are of great interest, as being probably the earliest relating to astronomy on record; and a summary of them will, therefore, not be out of place here. It must also be observed, that these paragraphs have each a commentary of far more recent date; without which, such is the abstruseness of their style, there would be great difficulty in understanding them.

In the first of these paragraphs Yaou is described as commanding He and Ho ' to observe the heavens, to compute the calendar, to form an instrument by which the motions of the Sun, Moon, and twelve signs might be represented, and with due respect to impart information respecting the seasons to the people.'

The comment on this paragraph is worthy of attention. In it we are informed that one of the words employed (Leih) signifies the recorded observations or computations. Another (Seang) refers to an instrument, probably resembling our armillary sphere. It is also stated that the Sun represents the male, or superior principle of nature, and the Moon the female, or inferior principle; that the Sun passes round the Earth in one day, and that the Moon is every lunation in conjunction with the Sun; that the word Sing, or stars, indicates not only the 28 stellar divisions, but also the five planets, Mercury, Venus, Mars, Jupiter, and Saturn; and the commentator fancifully compares the heavens to a piece of cloth in the loom, the stars forming the warp and the planets the woof: thus not inaptly indicating the paths of the planets among the fixed stars. Another word (Shin) is explained as signifying the twelve places in which the sun and moon are in conjunction: thus, in some measure, answering to our twelve signs. This may serve to give some notion of Chinese astronomy in those early times, and also to show the general nature of the commentary.

In the second paragraph Yaou establishes a division of the duties, and orders He Chung (or, as he may be called, He the Second) to go to a certain place in the East. He directs him to receive the rising Sun with due respect (that is, to perform the ceremonies necessary for that purpose), and to arrange the business of the spring. He was to observe whether the days and nights were at that time of equal length. A certain star (Neaou) is mentioned as being the correct indicator of the season, and certain tests are named as showing the middle of spring. There are the people going abroad on agricultural business and the pairing of birds and beasts.

The Commentator informs us, that ' after the completion of the Calendar a division of the duties took place, in order that certain observations might be made to verify the computations, lest inadvertently some error might have been introduced. These form the subject of this and the three succeeding paragraphs.' He also observes, ' Some suppose that these particular instructions were given to the second and third brothers of He and Ho, while others are of opinion that He and Ho are official denominations, and not the names of individuals, and that the others were assistants of different grades: which opinion is correct,' says he, ' cannot now be rightly ascertained.' The duties to be performed in this verification are distinctly named, and the star ' Neaou Bird ' is said to refer to a star in one of the seven stellar divisions of the southern quarter, denominated that of the ' Red Bird.' He also informs us, that ' by a Chinese astronomer named Tang Yih Hing the star Neaou is considered to be the same as the zodiacal division Shun Ho, ' the Quail-fire.' This star appears to be identical with α or Cor Hydræ, which is the central star of that division, and which is said to have culminated at sunset on the day of the vernal equinox in the time of Yaou.

Now if α Hydræ were observed culminating at sunset on the day mentioned, the Sun must have been in our sign Taurus, or in the Chinese division Maou, determined by the Pleiades; which was, consequently, then the equinoctial point. Reckoning from the middle of that constellation (the Pleiades), we find it may be roughly estimated as being, at the present time, rather more than 58 degrees from the equinoctial point—say 58 degrees. Now the precession of the equinoxes being at the rate of about a degree in 72 years, by multiplying 72 by 58 we obtain 4176 years as having elapsed since the time of Yaou to A.D. 1870; and 4176 less 1870 will be equal to 2306 B.C. as the date of the observation. It has been before shown, that the reign of Yaou commenced in the year 2356 B.C. He is said to have reigned 100 years, and 2356 less 2306, the number just found, will give the 50th year of that reign. This may be considered sufficiently near for a rough computation like the present, and thus a strong presumptive proof is again afforded of the veracity of Chinese history as recorded in the 'Shoo King.'

In the third paragraph Yaou directs He Shuh, or He the Third, to go to a place in the South. He is there to observe, with due ceremony, the length of the Sun's shadow, and thus to ascertain the middle of summer. Another star (Ho) is mentioned as indicating that period, and the tests are, the people still more actively engaged in agriculture, the moulting of birds, and the change of the fur in animals. This evidently refers to the observation of the summer solstice by means of the shadow of the gnomon. The star Ho, or Ta Ho, is the central one of the seven stellar divisions of the western quarter, that of the 'Azure Dragon,' and is identical with β in Scorpio.

The fourth paragraph contains the instructions to Ho Chung, or Ho the Second. He is directed to proceed to the West, and respectfully to escort the departing Sun. The days and nights are again equal. The star (Heu) is mentioned as the indicator of the season, and the tests are the people resting from their labours, the birds being well fledged, and the beasts having sleek coats. The star Heu is the central one of the seven stellar divisions comprised in the northern quarter, that of the 'Black Warrior,' and is identical with β Aquarii.

In the fifth paragraph Ho Shuh, or Ho the Third, is commanded to go to the North, to observe the northern changes. The day is then at the shortest, and the stars Maou are mentioned as those by which the winter solstice may be correctly ascertained. The tests are, the people keeping themselves within-doors and the birds and beasts having their winter covering of down and hair.

The stars Maou form the central one of the seven stellar divisions of the eastern quarter, that of the 'White Tiger,' and answer to the Pleiades. I may here observe, that the stars mentioned as the indicators of the seasons are about six hours of R. A. apart from each other: thus affording another presumptive proof of the accuracy of the early Chinese astronomical observations. The stellar divisions and the four quarters mentioned will be fully explained in a subsequent part of this work.

In the sixth paragraph Yaou thus addresses his astronomers:—'He and Ho, ye know that a year has 366 days. Fix the intercalary moons, regulate the hundred offices, and all things will prosper.'

The Commentator upon this paragraph informs us, that the year of 366 days mentioned by Yaou is that of the revolution of the heavens, and that the length of the solar year is 365¼ days. He minutely describes the various computations needed for ascertaining the exact length of the year, with many other particulars of interest, but which can hardly be entered into here.

Such is the substance of these curious notices of early Chinese astronomy, perhaps the most ancient on record. It must, however, be borne in mind, that the correctness of this account depends entirely upon the degree of credence to be given to the 'Shoo King.' Assuming its authenticity, of which there can be but little doubt, we find that at a very remote period, *i. e.* between two and three thousand years before the Christian era, the Chinese had made great progress in astronomy, that they were acquainted with the true length of the year, that they observed the equinoxes and solstices, that they had discovered the necessity of frequent intercalations of moons, or months, to keep the seasons in their true places, and were able to perform the computations requisite for that purpose ; together with many other facts, proving the high degree of knowledge of astronomy to which they had attained.

The second section of the 'Shoo King,' called 'Shun Teen,' is devoted to the actions of the Emperor Shun, the successor of Yaou. In this the following curious passage occurs :—'He examined the Tseuen Ke and the Yuh Hang, that the seven Ching might be duly regulated or observed.' The Tseuen Ke was the instrument before mentioned as resembling our armillary sphere; it is described as having been enriched with pearls : and the Yuh Hang appears to have been a kind of quadrant, having a jewelled tube fixed transversely. The seven Ching are the Sun, Moon, and five planets. There is a very full commentary upon this passage, occupying nearly four pages. The object of this examination by Shun is said to have been that he might ascertain whether the instruments were in order, so as to enable correct observations of the heavenly bodies to be made ; which observations were required in the computation of the Calendar. There are some curious passages in this Commentary relating to the theories of the heavens, and many other particulars, explaining the construction and use of the before-mentioned instruments. There is also a description of one made upon the ancient principles, about A.D. 450, in which the tube is said to have been 8 cubits in length and 1 inch in diameter. In this both these instruments were combined in one, and the tube being fixed to one of the circles of the sphere, which was movable, it could be turned about, and the positions of the Sun, Moon, and other heavenly bodies, could be ascertained by looking through it.

There are many other allusions to astronomy in this very ancient book, the 'Shoo King.' The eclipse described by me in the 'Monthly Notices,' vol. xxiii. p. 238, which occurred in the year 2158 B.C., is there recorded.

In other early books of the Chinese, astronomical notices occur. In the 'She King,' a collection of ancient poems, selected and arranged in their present form by the celebrated Confucius, comets and the stellar divisions are alluded to. In the 'Chun Tsew,' a work written by Confucius, the eclipses of which I have given an account in the 'Monthly Notices,' vol. xxiv. p. 39, are recorded. In the 'Tso Chuen,' another ancient historical work, there are many astronomical notices ; and in the 'Urh Ya,' a kind of dictionary of terms, even then considered of high antiquity, compiled during the Chow dynasty, *i. e.* between B.C. 1122 and 314, the twelve Kung, or zodiacal signs, and many of the stellar divisions, are mentioned. The great historical work, the 'She Ke,' which commences with Hwang Te, about 2650 B.C., and to which I am indebted for a large proportion of the observations of comets detailed in the subsequent pages of this volume, is highly deserving of notice. This truly great work was commenced by the historian Sze Ma Tseen. He brings the history of China down to the year 97 B.C., and it has been continued by a succession of historians to the end of the Ming dynasty, A.D. 1644. In this work certain sections are devoted exclusively to astronomy; and these, of course, in the present investigation, are the most important. In these, among other interesting matters, are to be found observations of the Sun, Moon, and five planets; occultations of stars; and notices of extraordinary appearances in the heavens, among which comets hold an important place.

Astronomical notices also occur in many other historical and scientific works, among which the accounts of comets in the celebrated 'Encyclopædia' of Ma Twan Lin must be particularly mentioned. It is only recently I have obtained a sight of this important work, for which I am indebted to the Rev. J. Summers, Professor of Chinese in King's College, London, who has kindly favoured me with the loan of the volume containing the cometary observations; and has thus enabled me to render my list far more complete, both as to details and number, than it otherwise would have been. Ma Twan Lin flourished during the later Sung dynasty, A.D. 960–1279. His laborious compilation of the Encyclopædia bearing his name is looked upon by the Chinese as one of the most extraordinary works ever produced by man. It is much admired by them for the immense amount of information it contains, and for the elegance and perspicuity of its style. The volume I have just referred to contains notices of comets from B.C. 613 to A.D. 1222, shortly after which date the author appears to have died. A Supplement, bringing the work down to A.D. 1644, has since been published, containing the cometary observations from the death of Ma Twan Lin to that date. Of this I had previously seen a copy, and made the necessary extracts.

The 'Tung Keen Kang Muh,' an abridgment of Chinese history from the earliest times to the end of the Yuen dynasty, A.D. 1367, in 100 volumes, is another work containing brief accounts of comets, some of which are not found in the 'She Ke.' It has been translated into French by M. Mailla.

Various works, professedly on astronomy, also occur, from one of which the Chinese Celestial Atlas, hereafter to be noticed, has been copied. In one of these works, printed in 1652, there is a list of 155 of the most important treatises on astronomy then existing in China. These afford another proof of the great attention paid by the Chinese to that science. It must, however, be observed, that astrology is almost universally coupled with astronomy by that people.

Such is a very brief summary of the state of astronomy among the Chinese. As we proceed, other portions of the subject will be touched upon and explained. It is chiefly from the works just mentioned, and more particularly from the 'She Ke' and the 'Encyclopædia' of Ma Twan Lin, that the observations of comets, that form the subject of the present compilation, have been derived; and it may be observed, that the materials thus collected consist of observations of comets made under the various dynasties from the period of the Chun Tsew, B.C. 613 to A.D. 1640: shortly after which time the Ming dynasty was subverted by the present reigning one, the Tsing.

They commence with B.C. 613, that being the year in which the cometary observations of Ma Twan Lin begin. The observations of comets earlier than this are not only very few, but are also so vague and unsatisfactory in their details, that it was thought advisable to omit them altogether.

The number of observations of comets thus brought together amounts to 373. Some of these may possibly be meteors, and may consequently be rejected on future revision. M. E. Biot, in the Supplement to the 'Connaissance des Temps' for 1846, has published a catalogue of comets observed in China under the following heads :—

Those from A.D. 1230 to 1649, of which he notices . . 94
Those from B.C. 134 to A.D. 1203 64
Those near oppositions of Halley's Comet 66
_____
Making a total of 224

It appears, therefore, that the list of cometary observations in the present work contains 149 more than Biot's catalogue.

The translation is as literal as the idiom of the two languages would allow, and every care has been taken to make it as accurate as possible. It must, however, be observed, that no attempt has been made to translate the names of the Chinese asterisms, as no useful purpose would be answered by it ; and to give the meaning of a few and not of the whole would tend to introduce confusion in the narrative. The original names have, therefore, been everywhere retained. It may also be remarked that the Chinese names are quite as fanciful as our own. Thus, Canopus is called Laou Jin, 'the Old Man ;' Arcturus, Ta Keo, 'the Great Horn ;' the seven bright stars in Ursa Major, Pih Tow, 'the Northern Measure ;' and the stellar division in which our constellation Gemini occurs is called Tsing, 'the Well.' These and other Chinese words will be found in the English version of the text untranslated : they are, however, in every instance, fully explained beneath the text. They have been so placed, not only for the convenience of classification, but also as enabling explanatory remarks to be introduced where necessary.

The manner in which these observations are recorded in the original is more or less explicit. In some we have merely the dynasty, emperor, year, and moon ; in others, the day and place of the heavens in which the comet was seen are added ; and in those which are the most fully described we have, in addition to the particulars before mentioned, the path of the comet through the heavens : comprising the stellar divisions in which it was seen, the asterisms through which it passed, and the stars near to which it approached ; together with the various days on which it was observed and the length of time it was visible, its colour, the length and direction of the tail, and other circumstances considered worthy of notice.

The description may, therefore, be considered as divided into two general heads ; the one chronological, the other astronomical. In the chronological part we have to ascertain all particulars respecting the dates of the dynasty, the emperor, the epoch and its year, the moon or month, and the day on which the comet appeared, and the days subsequently mentioned until its final disappearance. In the astronomical part we have, in like manner, to ascertain the stellar division in which the comet was first seen, and those through which it subsequently passed; and the various asterisms and stars mentioned as being in its path. To these must be added the description of the appearance of the comet, as regards colour, length of tail, &c.

For the first of these objects, viz. the ascertaining the various dates mentioned, it has been found necessary to construct several Tables. The first portion of these consists of a complete set of Chronological Tables, in which are to be found the succession of the Emperors from the earliest times, the dates of their accession to the throne, and the duration of their epochs and reigns, reduced to our reckoning. These Tables comprise the whole of the dynasties considered as regular by the Chinese, in their succession from the most remote period to the present time, with the names of the Emperors and of the epochs adopted by them. These are arranged in columns. The names of the Emperors and epochs are given in the original characters, with the pronunciation in English; together with the date of the commencement and duration of each epoch and reign. To these are added Tables of the Minor Dynasties, with the names of their princes and epochs as far as could be ascertained. The whole from original sources.

In forming these Tables, valuable assistance has been obtained from a chronological work compiled by the Japanese Prince of Mito, and published in Japan about 1863; in which not only is the chronology both of Japan and China given from the earliest times to A.D. 1860, but also the corresponding dates B.C. and A.D., according to our mode of expressing them. This work affords much valuable information, and deserves great praise for the perspicuity of its arrangement and the able manner in which it has been carried out. I need scarcely say the work is in Japanese; but the characters being the same as the Chinese, and as, although differing phonetically, they have precisely the same meaning, there was, therefore, no difficulty in making them out. The title in Chinese reads, 'Sin Chuen Neen Peaou,—A newly compiled Guide to Years,' or Chronology.

The word Epoch having been frequently used, it may be necessary to explain what is meant by that term. In this and the succeeding pages, the word Epoch is employed to designate the appellation of the years of the Emperor's reign. The term is not strictly correct, the Chinese equivalent being 'Neen Haou,—The Years' Name,' or designation; but it is the nearest I could adopt. It is now about 2000 years since it has been the custom of the Chinese Emperors to assume certain adulatory titles to express the years of their reign; and it is by these titles these personages are designated by the people at large and by strangers. The true name of the Emperor is never mentioned, as it would be considered as highly insulting to him to do so. Upon his death another name is given him, by which he is hereafter to be known in history. This is called his Temple name, being that placed in the Temple of Ancestors. It follows, therefore, that Kang He, Keen Lung, Taou Kwang, are not the names of the Emperors thus usually designated, but only the appellations of the years of their respective reigns; and in history they are only known as Shin Tsoo, Kaou Tsung, and Tseuen Tsung. It was formerly customary to change the epoch several times during a reign, and we have one instance in the early part of the Han dynasty of 11 such changes in a reign of 54 years; and under the Tang dynasty there are no fewer than 14 changes in a reign of 34 years. From the accession of the Ming dynasty, A.D. 1368, to the present, excepting in one instance, no change has been made in the epoch during the reigns of any of the

Emperors, that assumed at the accession having been kept until the close of the reign. These circumstances render the study of Chinese history a matter of some difficulty at the first, and hence the value of accurate tables in any investigations involving dates.

In using these Tables, the dynasty having been ascertained, the names of the Emperors of that dynasty and of their epochs, with their dates, will be found in their respective columns. For example: Required the 3rd year of the Epoch Woo Fung, of the Emperor Seuen Te, of the Western Han dynasty. On reference it will be found that Seuen Te was the eighth emperor of that dynasty; that he reigned 25 years, from B.C. 73–49; that Woo Fung was his 5th epoch, extending from B.C. 57–54: consequently its 3rd year was B.C. 55. Again, Tang dynasty: Required the 2nd year of the Epoch Han Hang, of the Emperor Kaou Tsung. On reference it will be found that Kaou Tsung was the third emperor of that dynasty, who reigned 34 years, from A.D. 650–683, and that Han Hang was his 7th epoch, from A.D. 670–673. The 2nd year of the Epoch Han Hang was, therefore, 671. It will be seen from these examples, that these Tables give all the information required for ascertaining the date of any year, according to our reckoning, that may occur in Chinese history.

Having thus ascertained the year, we have next to find the moon, or month, and the day of the year, on which a comet appeared, or any other remarkable circumstance occurred. For understanding the method of computing these, some acquaintance with the Chinese Calendar is required.

The Chinese year is luni-solar, and is reckoned by lunations, or moons as they term them; which may be considered as answering to our months, and of which 12 make up the ordinary year. These moons are of 29 or 30 days, regulated by certain fixed rules. They, however, are not alternate, and the common year consists of but 354 or 355 days. Hence the necessity of frequent intercalary moons at short intervals, there being seven of these moons in the cycle of 19 years, and consequently they fall generally between every second and third year. The year thus increased consists of 384 or 385 days; and in this manner the deficiencies of former years are made up, and the seasons kept in their proper places. This mode of intercalation appears to have been practised from extremely remote antiquity, as it is mentioned, as I have before shown, in the instructions of Yaou to his astronomers, more than 2000 years before the Christian era.

The succession of the moons in any one year is regulated by the first day of that year, which is not a fixed day, as with us, but, like our Easter Sunday, is not the same for two consecutive years. The first day of the Chinese year is the first day of the lunation in which the Sun enters our sign Pisces: it may, therefore, be any day between January 22 and February 20 inclusive. Hence it follows that this first day of the year must, of necessity, be ascertained before the moons can be properly appropriated. For this purpose the lunar cycle of 19 years must be employed; and a Table of the first year of each of these cycles, from B.C. 609 to A.D. 1995, has been constructed: as also another Table, showing the first day of each lunation in every year of the 19-year cycle. These Tables are formed from those in 'L'Art de Vérifier les Dates.' In order to use them, we must find in the first of these Tables the number of the given year in the 19-year cycle in which it occurs, and against that number in the second Table will be found approximately the first day of each lunation in that year. For example: Let it be

required to find the 1st day of the 6th moon in the year A.D. 678.   In the Table of the
first years of cycles, 684 is the nearest below that number, consequently 698 is the 15th
year of that cycle; and in the second Table it will be found that the 1st day of the 1st
moon in the 15th year of the cycle is February 17, and the 1st day of the 6th moon
July 15, the day required.   Again: Required the 1st day of the 10th moon, A.D. 1448.
Here 1444 is the 1st year of the cycle in which 1448 occurs, of which it is the 5th year,
the 1st moon of which commences February 7; and the 1st day of the 10th moon is
September 2.   It must, however, be observed, that these Tables must be considered as
approximate only: they are, however, sufficiently accurate for the purpose required.
It must also be remarked, that the earliest date on which the first day of the Chinese
year can fall is January 22; and whenever the second lunation in the Table commences
in February, after the 20th, the lunation commencing in January is to be taken as the
first of that year, and the succeeding moons reckoned accordingly.   Thus, in the 14th
year of the cycle of 19 years the lunations commence with January 30, February 28,
&c.: in this case January 30 is the first day of the Chinese year.   In the 11th year the
moons are January 3, February 2, &c.   Here the first day is February 2.

The mode of reducing Chinese days to our reckoning is the next point to be considered.   In order to comprehend this it is necessary, first, to explain the principles of
the system by which the Chinese arrange their chronology.   They reckon by means of
periods, or cycles, of 60 years; the years in these cycles being regulated by means of
the combinations of two series of characters, the one of 10 the other of 12.

The following Table shows these characters in the order in which they occur:—

FIRST SERIES, 10.		SECOND SERIES, 12.	
甲	Kea	子	Tsze
乙	Yih	丑	Chow
丙	Ping	寅	Yin
丁	Ting	卯	Maou
戊	Woo	辰	Shin
已	Ke	巳	Sze
庚	Kang	午	Woo
辛	Sin	未	We
壬	Jin	申	Shin
癸	Kwei	酉	Yew
		戌	Seuh
		亥	Hae

d

This system is called Kea Tsze, from the names of the first characters in each series. It is said to have been first introduced by the Emperor Hwang Te, the first year of the first cycle being reckoned as the 61st of that emperor's reign, answering to B.C. 2637. Whether this statement be correct or not this is certain, the system has been in use from extremely remote antiquity, and is employed in all their historical works, however early, to express the various dates that occur in them.*

They are employed thus:—The characters in the first series are combined with those in the second, from the first to the tenth, in this manner,—Kea Tsze, Yih Chow, &c. to Kwei Yew. The first character in the first series is now combined with the eleventh of the second, Kea Seuh; and the second of the first with the twelfth of the second, Yih Hae; and the other combinations follow in due order. Proceeding thus, after sixty combinations, the last being Kwei Hae, the first characters in both series come together again, and a fresh cycle commences, the combinations of the characters following in the same order as before. This system is employed not only to express the years of the cycle, but also months, days, and hours. It is also applied to the points of the compass, and any other expression of numbers in a series of ten or twelve.

The Chinese days of the year are not reckoned, as among us, by weeks of seven days, each day having a definite name, but by cycles of 60 days, the characters of which are the same as those of the cycle of 60 years. The names of the days also are the same as those of the combinations of the Kea Tsze.

The ordinary year consists of six of these cycles of 60 days, making 360 days; consequently they fall short of the true number of days in the year—in common years by 5 and in leap years by 6 days. Hence there is a continual shifting of the characters for any particular day. If, however, the characters for a certain day in any one of our years—say January 1, 1860—are known, the characters for any other day in that year are easily ascertained. The characters for the 1st of January in any year are to be found by means of a Table, whose construction I will now explain. I have just remarked, that the reckoning of the days of the year by periods of 60 days, according to the Chinese method, falls short of the true year by 5 days in common and by 6 days in leap years. Hence it follows, that in the cycles of 60 days the characters for the 1st of January in any year being known, those for the same day in the succeeding year will be five in advance; unless it should be leap year, when they will be six in advance. Let us assume the characters for the 1st of January, 1860, to be those of the first of the cycle, Kea Tsze; those for 1861 will be Ke Sze, the sixth combination; those for the same day in 1862 will be Kea Seuh, the eleventh combination; those for 1863, Ke Maou, the sixteenth; and those for 1864, a leap year, Yih Yew, the twenty-second: the first three being five in advance and the last six. Proceeding thus, taking every fifth combination for common years and every sixth for leap years, we shall find, after eighty combinations, on the eighty-first the first combination, Kea Tsze, will recur, followed by the succeeding ones in precisely the same order as before; and thus a

* The whole of the Tables referred to in this and the succeeding pages will be found in the Appendix.

general Table will be formed, showing the characters for the 1st of January for 80 years. In the Table the combinations are numbered from 1 to 80, for the convenience of reckoning. It must also be observed, that the Julian reckoning is that to be employed in reducing Chinese time.

In order to find by this Table the characters for the 1st of January in any given year, a second or auxiliary Table is required. In this the year of the commencement of each period of 80 years, from B.C. 2561–1920, is given. They are arranged under the letters B.C. and A.D. For years A.D. subtract from the given year the next lower number in this second Table, and against the number thus ascertained the characters for the 1st of January in that year will be found. A few examples will render this clear : —

Required the characters for January 1, A.D. 943.

943 — 880 (the next lower number in the second Table) = 63. Against No. 63 in the 80-year Table are Jin Yin, the characters required.

Required the characters for January 1, A.D. 1396.

1396 — 1360 (the next lower number) = 36; against which are Kang Shin, the characters required.

Required the characters for January 1, A.D. 1868.

1868 — 1840 = 28 ; against which are Woo Seuh, those required.

To exemplify the correctness of these results, I may observe that Gaubil informs us that the characters for January 1, A.D. 1267, were Kwei Hae.

1267 — 1200 = 67 ; against which are Kwei Hae.

And again, that those for January 1, A.D. 638, were Sin Yew.

638 — 560 = 78 ; against which are Sin Yew.

For years B.C. the process differs slightly. Here we have to subtract the given year from the next higher number, and proceed as before.

Required the characters for January 1, B.C. 643.

721 (the next higher number) — 643 = 78 ; against which are Sin Yew, the characters required.

Required the characters for January 1, B.C. 279.

321 — 279 = 42 ; against which are Jin Sze, those required.

To exemplify this, Idler informs us that the characters for January 1, B.C. 198, were Ting Sze.

241 — 198 = 43 ; against which are Ting Sze.

Such is the extremely simple method to be pursued to find the characters for our 1st of January in any given year, B.C. or A.D. To find the days mentioned in the account of any occurrence or phenomenon, such as the appearance of a comet, &c., we must return to the Table of 60 days.

It has already been shown, that the first combination in that Table recurs on the 61st, and commences a new cycle, either of years or days, as the case may be. Hence it is evident, that the characters for January 1 in any year must recur on the first day of each subsequent period of 60 days; and, therefore, that in common years the characters for March 2, May 1, June 30, August 29, October 28, and December 27, being the first days of each period, must be the same as those for January 1. In leap years they recur on March 1, April 30, June 29, August 28, October 27, and December 26. It follows, then, that the characters for January 1 in any year being known, those for any other day in the same year can be easily ascertained. For this we must proceed in the following manner :—Having by the methods before mentioned found the month according to our reckoning, answering to the Chinese moon in which the given day occurs, we must then ascertain within which of the dates just mentioned as those of the recurrence of the characters for January 1 it is to be found. Let us suppose the day required to be one in the moon answering to our month of July: it will then fall between June 30 and August 29. In this case June 30 assumes the characters for January 1 ; and now, by counting on from that combination in the Table of 60 days, commencing with its date, June 30, until we arrive at the characters of the day required, we obtain the date of that day. For example :—

Required the day Sin Chow, in the 7th moon, A.D. 365. We have first to find January 1, thus, $365 - 320 = 45$; against which we shall find in the 80-year Table Woo Shin [5], the characters for January 1. 365 is the 5th year of the lunar cycle, in which year the 7th moon commences August 1. The 60-day cycle, in which this date occurs, commences June 30, which is consequently Woo Shin [5]. Call this June 30, and count on to Sin Chow [38] in the 60-day Table, and the date will answer to August 2, which is that required. The small figures in brackets refer to those after the Chinese combinations of characters in the 80-year Table, and are their numbers in the 60-day Table. Thus, Woo Shin is the 5th and Sin Chow the 38th in that Table. These numbers greatly facilitate the finding the required characters in the 60-year Table.

The following example will, I trust, fully exemplify the nature of the computations requisite in reducing Chinese time to European reckoning. It is a copy of one of the observations of comets recorded in the subsequent part of this volume.

It is stated that during the Sung dynasty, in the reign of the Emperor Le Tsung, in the 5th year of the epoch King Ting, the 7th moon, on the day Kea Seuh, a comet appeared. It was also observed on the days Ke Maou, Sin Sze, Woo Woo, Kea Tsze, and Sin Wei, when it disappeared.

On reference to the Chronological Tables it will be found, that the Sung dynasty ruled China from A.D. 960–1279. Le Tsung was the fourteenth emperor of that dynasty, and reigned from 1225–1264. King Ting was his eighth epoch, 1260–1264, the fifth year of which was 1264, the year required. To find the characters for January 1 in that year: $1264 - 1200 = 64$, against which, in the 80-year Table, will be found Ting Wei [44], which are, consequently, the characters for January 1 : 1264 is the 11th year of a cycle of 19 years. The 7th moon in that year of the cycle commences towards the end of July, in which case the nearest preceding date on which the characters for January 1 recur is June 29, 1264 being a leap year. Now count on from Ting Wei [44], June 29, to Kea Seuh [11], which will be found to be July 31 ; thence to Sin Sze [18], August 2 ; thence to Ke Maou [16], September 8 ; to Kea Tsze [1], September 14 ; and to Sin Wei [8], September 21, on which day the comet disappeared.

Having thus explained the mode of reducing the various dates occurring in these observations to European reckoning, I pass on to the second, or Astronomical division of the subject; in which we have to consider the manner in which the place of the comet and its course among the stars are to be ascertained. For understanding this, it will be necessary to give a brief summary of some of the principles of Chinese astronomy.

The Chinese divide the visible heavens into 31 portions; 28 of these may be termed the stellar divisions, and receive their names from, or are determined by, an asterism, generally forming the central or principal one of the division. The determination by an asterism having the same name has been preferred by me to that by any particular star in that asterism, as being, to the best of my judgment, more in accordance with the Chinese mode of proceeding; in which, as far as my experience goes, the asterism alone is mentioned, and not a determining star in that asterism. Various other asterisms make up the remainder of the divisions. These divisions are very irregular in their extent, both from north to south and from east to west, no two being alike in these particulars; the largest extending north and south from Perseus to Argo, and east and west 32° 49', while the smallest consists only of the few small stars in the head of Orion and of some other small stars in the immediate neighbourhood, extending from east to west but 24'.

In the Appendix will be found a Table of the 28 stellar divisions, their determining asterisms, and their extent north and south, and east and west.

In addition to these divisions there are three large spaces, denominated Yuen; a word signifying a wall, or enclosure. These are, Tsze Wei Yuen, which may be considered as comprising stars within the circle of perpetual apparition; Teen She Yuen, consisting of stars contained within a line drawn through the constellation Serpens and continued to the circle of perpetual apparition: thus comprising the upper part of Ophiuchus, Hercules, Corona Borealis, and some stars in Boötes, Aquila, and Taurus Poniatowski. The third space is called Tae Wei Yuen: this is contained within a line drawn through β, γ, δ, ε and others in Virgo, and β, σ, ι, θ and δ Leonis, and continued, as in the preceding instance, to the circle of perpetual apparition; thus comprising stars in Virgo and Leo, Coma Berenices, and others in Canes Venatici, Ursa Major, and Leo Minor. It must, however, be observed, that in the cometary observations the 28 stellar divisions are frequently alluded to as extending to the Pole, without reference to these three spaces. Thus, in several instances, the comet is described as having passed through 12 or even 15 of these stellar divisions before it disappeared, all its early places having been within the circle of perpetual apparition; where such a circumstance might easily happen, on the assumption that the stellar divisions were continued to the Pole, without its course being in any way extraordinary, on account of its high northern latitude.

As these divisions are continually referred to in the astronomical observations of the Chinese, an acquaintance with them is essential in investigations such as form the object of this work. Tracings have, therefore, been made from original charts in a Chinese treatise on astronomy, so as to form a complete Celestial Atlas, fully elucidating their method of representing the heavens. This Atlas comprises the greater number of the

e

asterisms referred to in these observations. A few names, however, occur in them that are not to be found in any of the charts or lists I have hitherto met with, and are, consequently, mentioned as unascertained. The Atlas consists of maps of the 28 stellar divisions just referred to, with the names of the asterisms as they occur in the original map, and their pronunciation in English, with an account of the stars composing them according to our nomenclature. This Atlas will be found in the Appendix to this work.

In compiling the explanatory part relating to this Atlas, great assistance has been derived from a tract entitled 'Chinese Names of Stars and Constellations,' which forms an appendix to Morrison's Chinese Dictionary, and which was contributed to that work by the late John Reeves, Esq., formerly a Fellow of the Royal Astronomical Society. Another Catalogue, by Father Franciscus Noel, contained in his 'Observationes Mathematicæ et Physicæ in India et China facta' (4to. Prague, 1710), has also been found of great service, as corroborating Reeves or throwing light on doubtful cases. Nothing, however, has been taken for granted; the stars depicted in these maps having been carefully verified by reference to, and comparison with, other star-charts, both European and Chinese. To these is added an Index, by which, the name of the asterism being known, the chart in which it occurs can be readily found; and in order to render this Atlas still more intelligible, reduced drawings of the figures in Flamstead's Atlas have been made, and the principal Chinese asterisms laid down upon the corresponding stars in them.

The Chinese arrange these 28 stellar divisions under four general heads, answering to our east, west, north, and south. These divisions are of very remote antiquity, and have received the names of Tsing Lung, 'the Azure Dragon;' Heung Woo, 'the Black Warrior;' Choo Neaou, 'the Red Bird;' and Pih Hoo, 'the White Tiger.' Each of these comprises three of the divisions called Kung, answering to, although not identical with, our zodiacal signs. The nature of these Kung will shortly be explained. Under the first of the four above-mentioned divisions, the Azure Dragon, considered by the Chinese as the autumnal quarter, we have three of the Kung, answering to our signs, Libra, Scorpio, and Sagittarius; and seven of the stellar divisions, those from Keo to Ke (see Table of the 28 Stellar Divisions), comprising stars from Virgo to Sagittarius. Under the second of these, the Black Warrior, we have three Kung, answering to Capricornus, Aquarius, and Pisces; and seven stellar divisions, those from Tow to Peih, extending from stars in Sagittarius to others in Pegasus and Pisces. Under the White Tiger we have three Kung, answering to Aries, Taurus, and Gemini; and seven stellar divisions, from Kwei to Tsan, *i. e.* from stars in Andromeda and Pisces to those in Orion. Under the Red Bird three Kung, answering to Cancer, Leo, and Virgo; and seven stellar divisions, being those from Tsing to Chin, viz. from stars in Gemini to Corvus.

The Chinese divide their year into 24 portions, of 15 days each, thus making up the number of 360 days: these 24 portions are termed Tsze Ke, the particulars relating to which will be found in a Table in the Appendix.

Of these 24 divisions, twelve, called Kung Ke, or Kung only, require more particular notice, inasmuch as they mark the twelve places in which the Sun and Moon come into conjunction; and are thus, in some degree, analogous to our twelve signs of the Zodiac.

They are distinguished in the Table by an asterisk. But it must not be supposed that the ancient names of these are in any way identical with our names of the signs; neither must they be confounded with the appellations introduced by the Jesuit Missionaries when they reformed the astronomy of the Chinese. They then adopted a set of names closely agreeing with our nomenclature, such as the White Ram for Aries, the Golden Bull for Taurus, and so on. It has been supposed by some, that as these names, agreeing so closely with those employed by us, are in use among the Chinese, they afford a convincing proof of the immense antiquity of our designations of the zodiacal signs. But no traces of these recent names are to be found in Chinese astronomy as it existed before the accession of the present dynasty, and, consequently, all inferences as to their antiquity, deduced from the correspondence of the Chinese names of the zodiacal signs and those employed in European astronomy, are wholly untenable, as no such affinity between the two sets of names actually exists.

In the Appendix is a Table, showing the names of the ancient Chinese Kung Ke, placed side by side with the modern names, commencing with our sign Aries. These ancient names are extracted from the Astronomy of the Ming Dynasty, published at the commencement of the present dynasty, where they mark the divisions of a catalogue of stars; and the total want of correspondence between these and the names introduced by the Jesuit Missionaries is clearly demonstrated by this Table. These last names are taken from a modern Chinese work on astronomy published in 1819, in which the two sets of names occur side by side, and are thus distinguished: the ancient denominations are termed 'Chung Kwo Ming,' Middle Nation, or Chinese names; and the modern ones, 'Sze Kwo Ming,' Western Nation, or European names.

The 12 Kung are not only used by the Chinese in regulating the equinoxes, solstices, and lunations, but they are also employed in the computation of their Calendar, to ascertain the intercalary moons. It has already been stated, that these Kung Ke mark the places of the conjunctions of the Sun and Moon, and, consequently, those of the new moons, or lunations. Now as each Kung Ke indicates a period of 30 days, and a lunation is of but 29 days and a fraction, it follows that, sooner or later, two new moons must occur in one of these Kung, or periods of 30 days. Whenever this happens, that moon is an intercalary moon. In order to illustrate this, let us consider the upper line in the following diagram as representing a series of Kung periods of 30 days, and the lower one a series of new moons of 29 days and a fraction.

From this it is evident that after a time a lunation occurs as at A, that falls entirely within a Kung period which, consequently, has two new moons in it. This is the intercalary moon; and hence the Chinese rule, 'The intercalary moon is without a Kung.' Of these intercalary moons there are seven in the lunar cycle of 19 years.

The intercalary moon immediately follows the moon from which it receives its designation. Thus, on reference to a Chinese Almanac for the 7th year of Heen Fung, 1857, I find the intercalary 5th moon immediately following the regular 5th moon of the Calendar of that year.

The character 閏, Jun, by which the Chinese designate the intercalary moon, affords a striking instance of the figurative nature of many of the Chinese characters. It is a compound one, formed of 門, Mun, a gate, or entrance, and 王, Wang, an emperor. In each of the seasons it is the duty of the Emperor to officiate monthly in certain religious ceremonies, in halls provided for that purpose, which are arranged in a square, the sides facing the cardinal points. They call the building in which these halls are contained 'Ming Tang.' These ceremonies are fully particularised in the ' Le Ke,' the Book of Ceremonies, or Rites; and are to be found in that work in the sixth book, called ' Yue Ling.' From this we learn that the Emperor, in the spring, performs the rites proper for that season in that part of the building facing the east; that his dress and other appointments are of a certain colour (green); with many other particulars not necessary to be mentioned here. In the summer season the ceremonies are performed in the halls facing the south, the dress, &c. being of another colour; and so on for the remaining seasons. But there is no hall provided for the ceremonies required in the intercalary moon; they are consequently performed in the gateway, or entrance to the building: and hence the character Jun, representing the Emperor in the gateway, as that for the intercalary moon, is a very appropriate and significant symbol of this peculiarity in the performance of the rites for that moon. The institution of these ceremonies dates from extremely remote antiquity, and I may add that there is every appearance of their being still in use; for as late as 1787 the Emperor Keen Lung was, by a decree of the Tribunal of Rites and Ceremonies, allowed to perform these rites by deputy, his great age and consequent infirmities rendering it impossible for him to support the fatigue of going through them in person.

The 28 stellar divisions are evidently of very great antiquity, as the names of many of them occur in their most ancient works. They are to be found, together with the principal asterisms and stars composing them, in the Astronomical section of the Early Han Dynasty, in the 'She Ke,' which was first published in the first century of our era. This section also contains rules for forming the Calendar and computing the ordinary and intercalary moons, together with observations of the Sun, Moon, and Planets, and of extraordinary appearances in the heavens, among which those of comets occupy a prominent position.

In the Astronomical section of the Annals of the Tang Dynasty, A.D. 618–906, is an enumeration of the 28 stellar divisions, and the asterisms composing them; with notices of Eclipses and of the 12 Kung; and also observations of the Sun, Moon, Planets, Comets, &c.

The Astronomy of the Ming Dynasty, A.D. 1368–1644, is, as might be expected, much more expanded; embracing not only the whole of the before-mentioned particulars, but also comprising Tables of the Sun, Moon, and Planets, together with a Catalogue of Stars, with their latitudes and longitudes, both on the equator and the ecliptic. We are probably indebted to the Jesuit Missionaries for the greater part of this additional matter, as the Tables in particular bear evident marks of being from European sources.

A brief summary of the subjects treated upon in the 'Teen Wan,' or Astronomical section of the history of the Ming dynasty, and contained, according to my copy, in the 7th vol. of the History of that dynasty, will serve to give a more definite idea of the general nature of Chinese astronomy. It is divided into three chapters, the first of which has nine subdivisions, or sections. The first of these sections treats of the Leang E; that is, of the two great divisions of the universe, Heaven and Earth. In the second are notices of the Tseih, or seven Ching, which are enumerated as the Sun, Moon, and five Planets. The third section, Hang Sing, 'Perpetual Stars,' relates to the fixed stars. In this section is the Catalogue of Stars before referred to, consisting of 109 stars, with their degrees reckoned upon the equator and the ecliptic. In the fifth section the places of 16 of the stellar divisions, in degrees of the 12 Kung, or zodiacal signs, are enumerated in like manner. The sixth section relates apparently to the application of instruments to the observation of the heavenly bodies, with their mode of construction: the Tseuen Ke, or armillary sphere, and Yuh Hang, the Jewelled Tube, being particularly referred to. The seventh section is devoted to observations of the length of the shadow of the gnomon in various places, and in different seasons. The eighth relates to the method of reckoning the longitude; and the ninth to Chung Sing, 'Middle Stars,' by which term they appear to designate certain stars seen on the meridian at different seasons of the year.

The second chapter consists of four sections. The first of these is devoted to observations of Occultations of Planets by the Moon, and the following examples, showing their general style, may be of some interest. They commence thus:—

In the 1st year of the epoch Hung Woo, the 5th moon, on the day Kea Shin, Saturn was occulted (by the Moon): that is, on May 31, 1368.

In the 12th year of the same epoch, 3rd moon, day Woo Shin, Mercury was occulted: that is, March 13, 1380.

The second section relates to Occultations of Planets by each other. The observations run thus:—

Hung Woo, 6th year, 3rd moon, day Woo Shin, Mars occulted Saturn: that is, 1373, April 19.

In the 6th moon of the same year, day Jin Shin, Venus occulted Jupiter: that is, 1373, June 22.

The third section is entitled 'The Five Planets in one place,' by which conjunctions of several of the planets are evidently meant. The following are examples:—

Hung Woo, 14th year, 6th moon, day Kwei Wei, Mercury, Mars, and Venus were together in the stellar division Tsing: that is, 1381, May 22. These planets were in conjunction in Gemini.

In the 17th year, 6th moon, day Ping Seuh, Jupiter, Saturn, and Venus were together in the stellar division Tsan: that is, 1384, July 8. The stellar division Tsan is determined by the bright stars in Orion. The conjunction was most likely in Taurus or Gemini.

f

The fourth section treats of Stars Occulted by the Planets. The observations run thus : —

Hung Woo, 7th year, 8th moon, day Yih Sze, Jupiter occulted the great star in Heen Yuen (Regulus) : that is, on August 18, 1374.

The observations in this section are exceedingly numerous ; they occupy about 70 pages : but it is evident they are merely eye-observations, nothing like instrumental accuracy having been attempted ; and they are also to the nearest day only. Whether they are ever likely to be of any value to modern astronomers must be left to others to determine. They are exceedingly simple, and could be translated without the least difficulty. I may also observe that the word (Fan) which I have rendered ‘ occulted,’ signifies ‘ to screen,’ ‘ to shade,’ ‘ to put under shelter ;’ obviously implying our term, ‘ to occult.’

The next chapter contains nine sections. The first of these consists chiefly of stars seen in the daytime, being principally Venus, Jupiter, and Mars.

The next two sections are of much greater importance. They contain observations of what they term Kih Sing, or ‘ Temporary Stars,’ many of which are undoubtedly comets ; and of Suy Sing, ‘ Broom Stars,’ or comets. It is from these two sections most of the observations of comets recorded in the following pages as having been seen during this dynasty have been taken.

The next section records ‘ Changes in the Heavens ;’ and the succeeding one, ‘ Changes in the Sun and Moon,’ of which the following may be given as examples :—

Hung Woo, 2nd year, 12th moon, day Kea Tsze, a black spot was seen in the middle of the Sun : that is, January 1, 1370.

The same was observed in the 3rd year, 9th moon, day Woo Seuh ; 10th moon, day Ting Sze ; and 11th moon, day Kea Shin : that is, 1370, Oct. 2, Oct. 21, and Nov. 7.

The sixth section contains accounts of Haloes round the Sun and Moon ; the seventh, Changes in the Stars ; the eighth, Observations of Falling Stars ; and the ninth, accounts of extraordinary Clouds and Vapours.

The volumes which immediately follow the seventh contain, under another title, chiefly what we should perhaps call Meteorological Notices ; and those from the ninth to the twelfth inclusive are devoted to a collection of Tables of the Sun, Moon, and Planets, evidently from European sources.

I have already mentioned that I have preferred determining the stellar divisions by the asterisms which supply their names, instead of a particular star, as being more in accordance with the principles of the ancient astronomy of the Chinese. I may also observe, that in every instance, in the following Observations of Comets, where the stellar division is mentioned, the determining asterism alone is given. But as the rejection of particular determining stars takes away the points from which the computers of cometary orbits must start, it becomes desirable that the first degree of each stellar division, as given in original Chinese Charts, or Lists of Stars, should be ascertained as nearly as possible. Many of these have been carefully examined and collated, but I must express my regret that I have not hitherto met with any chart published before the introduction of the modern system ; all I have seen being comparatively of modern date, and commencing their degrees at the vernal equinox : whereas it appears to me most likely that the early Chinese astronomers, when their system of astronomy was first established, by placing the stellar division Keo (determined by *a*, &c. Virginis) first, in all probability commenced their reckoning with the autumnal divisions.

In order to supply the needful information as to the commencing degrees of the stellar divisions, I have been induced to form the following Table, which I trust will be found of service for that purpose.

No.	Name.	Degrees according to Chart.	Degrees according to Compass.	Determining Asterism.	Determining Star according to Biot.	First Degree of each S. D. according to Chart.
1	Keo	11	11	α Virginis and another	α Virginis	203
2	Kang	11	11	ι, κ, λ, θ Virginis	κ Virginis	213
3	Te	18	18	α, β, γ, ν Libræ	β Libræ	224
4	Fang	5	5	β, δ, π, ρ in Scorpio	π in Scorpio	242
5	Sin	7	8	α, σ, τ in Scorpio	σ in Scorpio	247
6	Wei	16	15	ε, μ, ν, &c. in Scorpio	μ²⁹ in Scorpio	254
7	Ke	9	9	ν, δ, ε, &c. Sagittarii	γ Sagittarii	270
8	Tow	24	24	ξ, τ, σ, φ, λ, μ Sagittarii	φ Sagittarii	279
9	New	8	8	α, β, &c. Capricorni	β Capricorni	303
10	Neu	12	11	ε, μ, ν, &c. Aquarii	ε Aquarii	311
11	Heu	10	10	β Aquarii and another	β Aquarii	323
12	Wei	20	20	α Aquarii, θ, ε Pegasi	α Aquarii	333
13	Shih	15	16	α, β Pegasi, &c.	α Pegasi	353
14	Peih	12	13	γ Pegasi, α Andromedæ	γ Pegasi	8
15	Kwei	12	11	β, δ, ε Andromedæ, &c.	β Andromedæ	22
16	Lew	13	13	α, β, γ Arietis	β Arietis	33
17	Wei	13	12	The three stars in Musca	α Muscæ	46
18	Maou	8	9	The Pleiades	η Pleiadum	59
19	Peih	15	15	α, γ, δ, ε, &c. Tauri	ε Tauri	67
20	Tsuy	1	1	λ and others in head of Orion	λ Orionis	82
21	Tsan	11	11	α, β, γ, δ, &c. Orionis	δ Orionis	83
22	Tsing	31	31	γ, ε, λ, μ, &c. Geminorum	μ Geminorum	94
23	Kwei	4	5	γ, δ, η, θ Cancri	θ Cancri	125
24	Lew	17	17	δ, ε, θ, &c. Hydræ	δ Hydræ	129
25	Sing	9	8	α, τ, &c. Hydræ	α Hydræ	146
26	Chang	18	18	κ, λ, μ, &c. Hydræ	ν³⁹ Hydræ	155
27	Yen	17	17	α, &c. Crateris	α Crateris	173
28	Chin	13	13	β, &c. Corvi	γ Corvi	190
		360	360			

The degrees in the preceding Table are taken chiefly from a Chart which appeared to me to be the most trustworthy of several which are in the possession of the Royal Astronomical Society. It consists of a planisphere bounded by a circle, on which the degrees are marked as on the equator. Lines meeting in the centre and cutting this circle indicate the extent of each stellar division to the nearest degree without fractions, the first degree marking the vernal equinox. The numbers are from 1 to 360, the first being the 8th degree of S. D. Shih, determined by *a* Pegasi and *a* Andromedæ. There are two dates on this chart: the earliest, possibly that of its construction, is the 25th year of Keen Lung, 1760; and the second, indicating its subsequent reproduction, the 13th year of Kea King, 1808.

Many other charts and authorities have been consulted. I may mention an exceedingly fine compass, formerly in the possession of the late Admiral Smyth, and presented to me by his widow, Mrs. Smyth. On it are twenty-four concentric circles, relating to the different purposes to which the compass is applied by the Chinese. Of these, two are devoted to the stellar divisions and their respective degrees; the one being of 360° the other of 365°. As there is a slight difference in the numbers of the chart and the compass I have in the preceding Table copied both.

It may also be necessary to observe that this Table contains the names of the 28 stellar divisions, with the degrees of each according to the before-mentioned chart and the compass circle of 360°; together with the stars composing their determining asterisms and the first degree of each stellar division. I have also introduced the determining stars of the stellar divisions according to Biot. And it may also be mentioned, that a line drawn from the centre through *a* Andromedæ cuts the fourth degree of the outer circle of the Chart. The other particulars that may be required to be known will be found in the Table of Stellar Divisions, and in the maps of those divisions in the Celestial Atlas, both of which form part of the Appendix.

The plan I have adopted in the translation of these Observations of Comets, in the MS. copy now in the library of the Royal Astronomical Society, is in every case to give the Chinese text in the original character, taken chiefly from the 'Encyclopædia' of Ma Twan Lin and the 'She Ke.' The reason of this is founded on the experience not only of its utility in a philological point of view, but also from its absolute necessity in any critical examination of the results; as without it no definite opinion can be formed as to the real import of the Chinese words employed: and it is much to be regretted that it was found impracticable to reproduce it in the present publication, on account of the extreme difficulty in procuring the necessary means of so doing. It may also be observed that as, in every instance, not only are the Chinese characters given, but also the corresponding sounds in English, according to Morrison's Dictionary, any character can without difficulty be found in that portion of the said Dictionary which is arranged according to the syllables; and thus any one so inclined can, with a very little application, verify for himself: and although he may have scarcely any, or even no knowledge whatever of the language, he can readily ascertain whether the ideas expressed in the translation are in accordance with the meanings of the characters as given in that Dictionary.

I may also remark, that in the translation the word 'Chih,' which is that by which the measure of the length of the tail of a comet is expressed, is everywhere rendered by 'cubit,' instead of 'degree,' the word used by Biot and others. As I had resolved to make my version as literal as possible, I could not consistently express that word by any other term. It is evidently used by the Chinese to express the length of the tail of a comet in an indefinite sense, just as we should employ foot or yard for that purpose; and the estimated length is consequently of about the same value. Readers may, however, if they think it advisable, substitute the word 'degree' for 'cubit;' but it must be borne in mind that that word does not express the Chinese idea, and, consequently, cannot be depended upon any more than the other, as giving the exact length of the tail of a comet. The word in Chinese for degree is 'Too;' this is a definite measure, but I do not find it anywhere employed in these observations to express the length of the tail of a comet, the word Chih, 'cubit,' being invariably used. I may also observe, that I have found no instance of the word Chih being used to express a 'degree.'

The 'Tsun,' or tenth part of the cubit, appears to be its unit. The original Tsun, or that of the ancients, is said to have been formed by placing ten grains of a certain cereal resembling our millet side by side, these seeds being of an oval form and pointed. The modern Tsun is estimated by placing ten of these seeds end to end, and thus there is a considerable difference between the ancient and modern Chih. The estimating the length of the Tsun by seeds is remarkable, as closely resembling our 'three barleycorns out of the middle of the ear' to make one inch. The relation of the measures of length that occur in the text is as follows:—

10 Tsun make 1 Chih,
10 Chih make 1 Chang.

From what I can ascertain the Chih is rather more than an English foot, the Tsun being about an inch and a fraction.

From the preceding remarks it must be evident, that the production of this work has been attended with no ordinary amount of labour. Many thousands of Chinese characters required to be carefully copied and accurately translated, the whole of the dates ascertained by computation, and numerous works, both Chinese and European, had to be examined or collated. In addition to these, the construction of the Tables for computing the dates of their chronology, and of the Atlas, both of which have been found not merely useful but indispensable to the carrying on of the work, required a great amount of research and attention. How far the results may be worthy of the time and labour bestowed upon them, must be left for those who are better qualified than myself to form an opinion on such subjects, to determine. Errors may doubtless be found to exist, although every care has been taken to avoid them; and it is hoped that none seriously affecting the character of any part of the work will be found. It must, however, be remembered, that this is strictly a work of reference, and as such may, at some future period, be of service in investigations respecting the former appearance of any particular comet that may then pay us a visit.

I have already mentioned various Chinese works employed in this computation. In

g

addition to these I must observe, that I have received much valuable information from works by European authors whose attention has been directed to Chinese astronomy and chronology. Among these I may mention Gaubil, whose 'Traité de la Chronologie Chinoise' has been of great assistance. A paper by Ideler, in the 'Abhandlungen' of the Berlin Academy for 1837, entitled 'Über die Zeitrichnung der Chinesen,' has afforded much valuable information. Pingré's 'Cométographie,' and J. B. Biot's 'Précis de l'Astronomie Chinoise,' have also been consulted with advantage; and Ed. Biot's 'Catalogue des Comètes observées en Chine,' published in the 'Connaissance des Temps' for 1846, has been carefully examined and collated; and I have much pleasure in testifying to the general accuracy of that work. Morrison's 'View of China,' for philological purposes, has been found of great service, as affording much miscellaneous information.

I must also express my acknowledgments to the Rev. Professor Summers, for his kindness in supplying me with Ma Twan Lin's 'Observations of Comets,' which have been found of the greatest value, as affording information not to be met with readily, if at all, elsewhere. And also for his valuable assistance in looking out and supplying me with such Chinese type as was required in the subsequent part of this work, which has enabled me to present it in a more efficient form than I could have adopted otherwise.

In conclusion I have only to express my confident expectation, that in placing the MS. of this work in the library of the Royal Astronomical Society, it will be in the most likely position to be of service to future inquirers into the subject of Chinese Astronomy, and more particularly of that portion of it which relates to their Cometary observations.

JOHN WILLIAMS.

April, 1871.

NOTE. — It may be necessary to offer some explanation of the departure from strict chronological order in the following Index to the Cometary Observations. Thus, in p. xxxii. after No. 269, A.D. 1264, Nos. 270, A.D. 941, to 277, A.D. 1237, follow. These are observations made in another part of China by other Astronomers during two contemporaneous minor dynasties—the Leaou, A.D. 916–1125, and the Kin, 1118–1236; and as I have strictly followed the arrangement of Ma Twan Lin, the above is the place in which they occur in his work. Some of these observations refer to comets noticed in the preceding accounts. Again, in the last column a number of observations, commencing in 1376, follow those ending in 1640. These are observations of the Kih Sing, or extraordinary stars, which form a separate section in the Astronomy of the Ming dynasty. It may also be noticed, that in A.D. 837 many comets are recorded as having appeared. Two undoubtedly, and possibly a third of these observations, refer to comets previously observed in the same year. They are given as they occur in 'M. T. L.'

INDEX TO THE SUCCEEDING COMETARY OBSERVATIONS.

No.	Year B.C.	Month & Day.	No.	Year A.D.	Month & Day.	No.	Year A.D.	Month & Day.
1	611	July	61	66	February 20	122	304	May
2	531	—	62	71	March 6	123	305	September
3	516	July	63	75	July 14	124	305	November 21
4	502	December	64	76	August 9	125	329	August
5	467	—	65	77	January 23	126	336	February 16
6	433	—	66	84	May 25	127	340	March 5
7	305	—	67	102	January 7	128	343	December 8
8	303	—	68	110	January	129	349	November 23
9	296	—	69	131	—	130	358	July 1
10	240	—	70	141	March 27	131	363	August
11	238	April	71	149	October 19	132	369	March
12	234	January	72	161	June 14	133	373	March 9
13	214	—	73	178	September	134	386	April
14	233	—	74	180	Winter	135	390	August 22
15	204	August	75	182	August	136	393	March
16	172	—	76	185	December 7	137	400	March 19
17	157	October	77	188	March	138	401	January 2
18	154	January	78	188	July 29	139	402	November 12
19	155	July	79	192	October	140	415	June 24
20	154	February	80	193	November	141	418	September 15
21	148	May	81	200	November 7	142	419	February 7
22	147	March 14	82	204	December	143	422	March 21
23	147	August 6	83	206	February	144	422	December 17
24	147	October	84	207	November 10	145	423	February 13
25	138	March	85	213	January	146	423	October 15
26	138	May	86	218	April	147	442	November 1
27	138	August	87	236	November	148	449	November 11
28	137	October	88	222	November 4	149	451	May 17
29	135	July	89	225	December 9	150	501	February 13
30	135	September	90	232	December 4	151	501	April 14
31	134	June	91	236	November 30	152	532	January 6 (?)
32	120	—	92	238	September	153	539	November 17
33	119	May	93	238	November 29	154	560	October 4
34	110	—	94	240	November 5	155	565	July 23
35	108 or 9	—	95	245	September 18	156	568	August 3
36	87	August	96	247	January 16	157	575	April 27
37	84	March	97	248	April	158	416	January 26
38	77	September	98	251	December 21	159	416	June 27
39	76	May	99	252	March 25	160	565	April 21
40	74	March	100	253	December	161	565	July 24
41	73	May 10	101	255	February	162	568	July
42	72	August 20	102	257	December	163	568	August
43	70	August 4	103	259	November 23	164	561	September 26
44	69	February	104	262	December 2	165	565	July 22
45	61	July	105	265	June	166	568	July 21
46	49	April	106	268	February 18	167	574	April 4
47	48	April	107	275	January	168	574	May 31
48	47	June	108	276	June 24	169	588	November 22
49	44	—	109	277	February	170	594	November 10
50	32	February	110	279	April	171	607	March 13
51	12	August 26	111	281	September	172	607	April 4
52	5	March 5	112	281	December	173	615	July
53	4	April	113	283	April 22	174	616	July
	A.D.		114	287	September	175	616	October
54	13	December	115	290	May	176	626	March 26
55	22	November	116	296	May	177	634	September 22
56	39	March 13	117	300	April	178	639	—
57	55	June 4	118	301	January	179	641	August 1
58	60	August 9	119	301	May	180	663	September 29
59	61	September 27	120	302	May	181	667	May 24
60	65	June 4	121	303	April	182	676	January 3

No.	Year A.D.	Month & Day.	No.	Year A.D.	Month & Day.	No.	Year A.D.	Month & Day.
183	676	July 7	246	1036	January 15	310	1452	March 21
184	681	October 17	247	1049	March 10	311	1456	May 27
185	683	April 20	248	1056	August	312	1457	January 14
186	684	July 8	249	1066	April 2	313	1457	June 15
187	684	September 13	250	1075	November 17	314	1457	October 26
188	707	November 16	251	1080	August 10	315	1461	August 5
189	708	March 30	252	1097	October 6	316	1465	March
190	708	September 21	253	1106	February 10	317	1468	September 18
191	710–713	—	254	1110	May 29	318	1472	January 16
192	730	June 30	255	1126	May 20	319	1490	December 31
193	739	March 27	256	1126	December	320	1500	May 8
194	760	May 16	257	1131	September	321	1506	July 31
195	760	May 20	258	1132	January 5	322	1506	August 10
196	767	January 12	259	1132	August 14	323	1520	February
197	770	June 15	260	1145	April 26	324	1523	July
197*	773	January 17	261	1145	June 4	325	1531	August 5
198	815	April	262	1147	January 6	326	1532	September 2
199	817	February 17	263	1147	February 12	327	1533	July 1
200	821	February 27	264	1151	August 21	328	1539	April 30
201	821	March 7	265	1222	September 15	329	1554	June 23
202	828	July 5	266	1232	October 18	330	1556	March 1
203	829	December	267	1240	January 31	331	1557	October 10
204	834	October 9	268	1240	February 23	332	1569	November 9
205	837	March 22	269	1264	July 26	333	1577	November 14
206	837	April 29	270	941	August 9	334	1580	October 1
207	837	May 3	271	1014	February 10	335	1582	May 20
208	837	May 21	272	1066	April	336	1585	October 3
209	837	June 17	273	1080	January 6	337	1591	April 3
210	837	June 26	274	1097	December 6	338	1593	July 20
211	837	September 9	275	1133	September 29	339	1596	July 26
212	838	November 11	276	1226	September 13	340	1607	September 11
213	838	November 21	277	1237	September 21	341	1618	November 16
214	839	February 7	278	1264	July 26	342	1619	February
215	839	March 12	279	1277	March 9	343	1639	Autumn
216	840	March 20	280	1293	November 7	344	1640	December 12
217	840	December 3	281	1299	June 24	345	1376	June 22
218	841	July	282	1301	September 16	346	1378	September 26
219	841	December 22	283	1304	February 3	347	1385	October 23
220	851	April	284	1313	April 13	348	1388	March 29
221	856	September 27	285	1315	November 28	349	1430	September 9
222	864	June 21	286	1337	May 4	350	1430	November 14
223	868	February	287	1337	June 26	351	1431	January 3
224	869	September	288	1340	March 24	352	1453	January 4
225	877	June	289	1351	November 24	353	1458	December 24
226	885	—	290	1356	September 21	354	1461	June 29
227	886	June 13	291	1360	March 12	355	1462	June 29
228	891	May 12	292	1362	March 5	356	1491	January 19
229	892	December	293	1362	June 29	357	1495	January 7
230	893	May 6	294	1363	March 16	358	1499	August 16
231	894	February	295	1366	October 25	359	1502	November 28
232	905	May 22	296	1368	February 7	360	1521	February 7
233	912	May 13	297	1368	April 8	361	1529	February 5
234	928	October 14	298	1373	May	362	1532	March 9
235	936	October 28	299	1391	May 23	363	1534	June 12
236	941	September 18	300	1407	December 14	364	1536	March 24
237	943	November 5	301	1431	May 15	365	1545	December 26
238	956	March 13	302	1432	February 3	366	1578	February 22
239	975	April	303	1432	Feb. 29 or Oct. 26	367	1584	July 1
240	975	August 3	304	1433	September 15	368	1604	September 30
241	989	August 13	305	1439	March 25	369	1609	—
242	998	February 23	306	1439	July 12	370	1618	November 24
243	1003	December 23	307	1444	August 6	371	1618	December 5
244	1018	August 4	308	1449	December 20	372	1621	May 12
245	1035	September 15	309	1450	January 19			

No. 197* makes up the full number, 373.

COMETS OBSERVED IN CHINA.

---

1   B.C. 611. *July.*

DURING the period of the Chun Tsew, in the 14th year of the reign of Wan Kung, Prince of Loo, in the autumn, in the 7th moon, a comet entered into Pih Tow.

The 'Chun Tsew' is a celebrated historical work, said to have been written by Confucius. It embraces the period between B.C. 722 and 481, and records the history of the princes of Loo, one of the minor states into which China was divided during the Chow dynasty. It was the native place of Confucius, and that in which he passed the greater portion of his life. In that work we are informed that the 14th year of Wan Kung corresponded with the 2nd year of the Emperor Kwang Wang, of the Chow dynasty, whose reign commenced B.C. 612. Hence the 14th year of Wan Kung was B.C. 611: 7th moon, July.

Pih Tow, the seven bright stars in Ursa Major.   *M. T. L.*

---

2   B.C. 531.

In the winter of the 10th year of Chaou Kung, Prince of Loo, there was a comet to the left of Ta Shin. It extended to Han.

Chaou Kung, B.C. 531: 10th year.

Ta Shin. According to the Commentary this appears to be a star in one of the stellar divisions, Fang Sin or Wei, all of which are determined by stars in Scorpio. The conclusion seems to be, that Ta Shin is Antares.

Han, possibly Teen Han, the Milky Way.   *M. T. L.*

---

3   B.C. 516. *July.*

In the 26th year of the same Prince, in the 6th moon, there was a comet near the star Tsze.

Chaou Kung, B.C. 516: 26th year, 6th moon, July.

Star Tsze, H Herculis.   *M. T. L.*

---

4   B.C. 502. *December.*

In the 13th year of Gae Kung, in the winter, the 11th moon, there was a comet in the east.

Gae Kung, B.C. 502: 13th year, 11th moon, December.   *M. T. L.*

B

CHOW DYNASTY, B.C. 1122–314.

5   B.C. 467.

In the 2nd year of the Emperor Ching Ting Wang a comet was seen.

     Emperor Ching Ting Wang, B.C. 468–441 : 2nd year, 467.   *M. T. L.*

6   B.C. 433.

In the 8th year of the Emperor Kaou Wang a comet was seen.

     Emperor Kaou Wang, B.C. 440–424 : 8th year, 433.   *M. T. L.*

7   B.C. 305.

In the 10th year of the Emperor Nan Wang a comet was seen.

     Emperor Nan Wang, B.C. 314–254 : 10th year, 305.   *M. T. L.*

8   B.C. 303.

In the 12th year of the same Emperor a comet was seen.

     Emperor Nan Wang, B.C. 303 : 12th year.   *M. T. L.*

9   B.C. 296.

In his 19th year a comet was seen.

     Nan Wang, B.C. 296 : 19th year.   *M. T. L.*

TSIN DYNASTY, B.C. 220–203.

10   B.C. 240.

In the 7th year of the reign of Che Hwang a comet first appeared in the east. It was afterwards seen in the north. In the 5th moon it was seen for 16 days in the west.

     Che Hwang was the Emperor who is said to have caused all the books to be burned and the literati to be destroyed. This was done in order that he might be considered by posterity as the founder of the Chinese Empire. His reign over the Tsin, one of the minor states, commenced B.C. 246. It was not until the 26th year he obtained the supreme power, and thus founded the Tsin dynasty. His reign is reckoned from 246; hence his 7th year is B.C. 240: 5th moon, May.

                                                  M. T. L.

11   B.C. 238. *April.*

In the 9th year of the same Emperor a doubtful star was seen in the horizon. In the 4th moon it was seen in the west. It was also seen in the north, to the south of Tow, for 80 days.

     Che Hwang, B.C. 238 : 9th year, 4th moon, April.

     Tow, most likely Pih Tow, the seven bright stars in Ursa Major.   *M. T. L.*

12 B.C. 234. *January.*

In the 13th year of the same Emperor, in the 1st moon, a comet was seen in the east.

Che Hwang, B.C. 234 : 13th year, 1st moon, January. *M. T. L.*

---

13 B.C. 214.

In his 33rd year a bright star appeared in the east.

Che Hwang, B.C. 214 : 33rd year. *M. T. L.*

---

14 B.C. 233.

In the Astronomy of the Han dynasty it is recorded, that in the time of Che Hwang, of the Tsin dynasty, in his 15th year, four comets were seen during 80 days. They extended to the horizon.

Che Hwang, B.C. 233 : 15th year. *M. T. L.*

HAN DYNASTY, B.C. 206 *to* A.D. 264.

15 B.C. 204. *August.*

In the 3rd year of the Han Emperor, the 7th moon, there was a comet near Ta Keo. *She Ke.*

In addition, 'Ma Twan Lin' gives the name of the Emperor, Kaou Te.
Kaou Te, B.C. 206–195 : 3rd year, 7th moon, August, 204.
Ta Keo, Arcturus.

16 B.C. 172.

In the 8th year of the Emperor Wan Te a tailed star appeared in the east.

Emperor Wan Te, B.C. 179–157 : 8th year, 172. *She Ke.*

17 B.C. 157. *October.*

In the reign of the same Emperor, the 7th year of the epoch How Yuen, the 9th moon, a comet appeared in the west. From first to last it was in Stellar Divisions Wei and Ke. It pointed towards S. D. Heu and Wei. It was about 10 cubits in length, and extended to Teen Han. After 16 days it was no longer seen.

Wan Te, epoch How Yuen (the first of the epochs), B.C. 163–157 : 7th year, 157, 9th moon, October.
Stellar Division Wei determined by ε, μ, ν, &c. in Scorpio.
Ke determined by γ, δ, ε, &c. Sagittarii.
Heu determined by β Aquarii and another.
Wei determined by a Aquarii and θ, ε Pegasi.
Teen Han, the Milky Way.

It must be observed here, that in the list of the S. D. there are no fewer than four whose names are expressed in English characters by Wei : of these, two occur in the description of this comet. The original characters of the whole of these are totally unlike.   *M. T. L.*

18   B.C. 154. *January.*

In the 2nd year of the Emperor King Te there was a comet in the south-west.

> Emperor King Te, B.C. 156–141 : 2nd year, 155, 12th moon, 154, January. There was no epoch for the first seven years of King Te.

19   B.C. 155. *July.*

In the 6th moon of the same year a comet appeared in the north-east.

> 6th moon, 155, July.   *She Ke.*

20   B.C. 154. *February.*

In the 3rd year, 1st moon, a tailed star was seen in the west.

> King Te, B.C. 154 : 3rd year, 1st moon, February.   *She Ke.*

21   B.C. 148. *May.*

In the 2nd year of the epoch Chung Yuen, the 4th moon, there was a comet in the north-west.

> Epoch Chung Yuen, B.C. 149–144 : 2nd year, 148, 4th moon, May. Chung Yuen was the first epoch of King Te.   *She Ke.*

22   B.C. 147. *March 14.*

In the 3rd year of the epoch Chung Yuen, the 3rd moon, on the day Ting Yew, a comet was seen at night in the north-west : its colour was white. It was 10 cubits in length. Its place was in Tsuy He. As it passed on it increased but little in size. After 15 days it was no longer seen.

> Chung Yuen, B.C. 147 : 3rd year, 3rd moon, day Ting Yew, March 14. Tsuy He, possibly S. D. Tsuy, λ and others in head of Orion.
>   *She Ke, M. T. L.*

23   B.C. 147. *August 6.*

In the 3rd year of the epoch Chung Yuen, the 6th moon, on the day Jin Seuh, there was a comet in the south-west : it was in the southern part of S. D. Fang. When it left Fang it was 20 cubits in length. It was as large as a two-tow vessel. Its colour was white. On the day Kwei Hae its place was to the north-east of S. D. Sin. Its length was then 10 cubits. On the day Kea Tsze it was in S. D. Wei. On the day Ting Maou it entered S. D. Ke, to the north, near the star Han. It gradually lessened, until it resembled a peach. On the day Tin Shin it disappeared, having been visible altogether for 10 days.

Epoch Chung Yuen, B.C. 147: 1st year, 6th moon, day Jin Seuh, August 6th; day Kwei Hae, August 7th; Kea Tsze, August 8th; Ting Maou, August 11th; Tin Shin, August 16th.

S. D. Sin, determined by Antares and others in Scorpio.

Wei determined by ε, μ, ν, and others, in Scorpio.

Ke determined by γ, δ, ε Sagittarii and others.

Star Han, ζ Ophiuchi. *She Ke.*

24   B.C. 147. *October.*

In the 9th moon of the same year there was a comet in the north-west.

Chung Yuen, B.C. 147: 9th moon, October.

25   B.C. 138. *March.*

In the reign of the Emperor Woo Te, the 3rd year of the epoch Keen Yuen, the 2nd moon, there was a comet in S. D. Chang. It passed through Tae Wei into Tsze Kung. It went into Teen Han. The 'Chun Tsew' says the comet was in Pih Tow.

Emperor Woo Te, B.C. 140–87; epoch Keen Yuen, 140–135: B.C. 138, 3rd year, 2nd moon, March.

S. D. Chang determined by κ, λ, μ, ν, φ Hydræ.

Tae Wei, space between stars in Leo and Virgo.

Sze Kung, circle of perpetual apparition.

Teen Han, the Milky Way.

Pih Tow, bright stars in Ursa Major. *M. T. L.*

26   B.C. 138. *May.*

In the 4th moon of the same year there was a comet in Teen Ke: it passed into Chih Neu.

Epoch Keen Yuen, B.C. 138: 3rd year, 4th moon, May.

Teen Ke, θ and others in Hercules.

Chih Neu, a and two other stars in Lyra. *M. T. L.*

27   B.C. 138. *August.*

In the 7th moon of the same year there was a comet in the north-east.

B.C. 138: 7th moon, August. *M. T. L.*

28   B.C. 137. *October.*

In the 4th year of the same epoch, 9th moon, there was a comet in the north-east.

Keen Yuen, B.C. 137: 4th year, 9th moon, October. *M. T. L.*

29   B.C. 135. *July.*

In the 6th year of the same epoch, the 6th moon, there was a comet in the west.

    Keen Yuen, B.C. 135 : 6th year, 6th moon, July.   *M. T. L.*

30   B.C. 135. *September.*

In the 8th moon of the same year there was a comet in the east : its tail extended across the heavens.   It was visible for 30 days.

    Keen Yuen, B.C. 135 : 6th year, 8th moon, September.   *M. T. L.*

31   B.C. 134. *June.*

In the 1st year of the epoch Yuen Kwang, the 6th moon, a strange star was seen in S. D. Fang.

    Epoch Yuen Kwang, B.C. 134–129 : 1st year, 6th moon, 134, June.
    S. D. Fang determined by β, δ, π, and others, in Scorpio.   *She Ke.*

32   B.C. 120.

In the 3rd year of the epoch Yuen Show, in the spring, there was a comet in the East.

    Epoch Yuen Show, B.C. 112–117 : 3rd year, 120.   *She Ke.*

33   B.C. 119. *May.*

In the 4th year of the same epoch, in the 4th moon, a comet appeared in the north-west.

    Yuen Show, 4th year, B.C. 119 : 4th moon, May.   *M. T. L.*

34   B.C. 110.

In the 1st year of the epoch Yuen Fung, the 5th moon, there was a comet in the eastern part of the S. D. Tsing.   It was in San Tae.

    Epoch Yuen Fung, B.C. 110–105 : 1st year, 5th moon, 110.
    S. D. Tsing, γ, ε, μ, ν, and others, in Gemini.
    San Tae, the feet of Ursa Major.   *M. T. L. Tung Keen.*

35   B.C. 109 *or* 108.

In the middle of the epoch Yuen Fung there was a comet in Hoo Shoo.

    Yuen Fung, middle B.C. 109 or 108.
    Ho Shoo, unascertained.   *M. T. L. Tung Keen.*

36   B.C. 87.   *August.*

In the 2nd year of the epoch How Yuen, the 7th moon, there was a comet in the east.

     How Yuen, B.C. 88–87 : 2nd year, 7th moon, 87, August.   *Tung Keen.*

37   B.C. 84.   *March.*

In the reign of the Emperor Chaou Te, the 3rd year of the epoch Che Yuen, the 2nd moon, there was a comet in the north-west.

     Emperor Chaou Te, B.C. 86–74.; epoch Che Yuen, 86–81 : 3rd year, 2nd moon, 84, March.   *Tung Keen.*

38   B.C. 77.   *September.*

In the 4th year of the epoch Yuen Fung, the 9th moon, there was a strange star in the middle of Tsze Kung. It was between the stars Choo in Tow and Keih.

     Epoch Yuen Fung, 80–75 : 4th year, 9th moon, 77, September.
     Tsze Kung, circle of perpetual apparition.
     Tow, the seven stars in Ursa Major.
     Choo, a Ursæ Majoris.
     Keih, Polaris.   *She Ke.*

39   B.C. 76.   *May.*

In the 5th year of the same epoch, the 4th moon, a bright star was seen between the S. D. Kwei and Low.

     Yuen Fung, B.C. 76 : 5th year, 4th moon, May.
     S. D. Kwei determined by β, δ, ϵ, and others in Andromeda, and stars in Pisces.
     Low determined by a, β, γ Arietis.   *She Ke.*

40   B.C. 74.   *March.*

In the 1st year of the epoch Yuen Ping, the 2nd moon, there was a large falling star like the moon. Many stars followed, all going to the west.

     Epoch Yuen Ping, B.C. 74 : 2nd moon, March.
     This appears to have been a large meteor.   *Tung Keen.*

41   B.C. 73.   *May* 10.

In the reign of Seuen Te, the 1st year of the epoch Pun Che, the 4th moon, on the day Jin Seuh, early in the evening, a tailed star appeared to the west of the S. D. Tsan.

     Emperor Seuen Te, B.C. 73–49 ; epoch Pun Che, 73–70 : 1st year, 4th moon, day Jin Seuh, 73, May 10.
     S. D. Tsan, determined by a, β, γ, δ, &c. Orionis.   *She Ke.*

42   B.C. 72. *August* 20.

In the 2nd year of the same epoch, the 7th moon, on the day Sin Hae, a comet appeared in S. D. Yih.

> Pun Che, B.C. 72 : 2nd year, 7th moon, day Sin Hae, August 20.
> S. D. Yih, determined by α and other stars in Crater.   *She Ke.*

43   B.C. 70. *August* 4.

In the 4th year of the same epoch, the 7th moon, day Kea Shin, a comet appeared in S. D. Yih : it passed near the moon.

> Pun Che, B.C. 70: 4th year, 7th moon, day Kea Shin, August 4.
> S. D. Yih, determined by α and others in Crater.   *She Ke.*

44   B.C. 69. *February.*

In the 1st year of the epoch Te Tsae, the 1st moon, there was a comet in the west.

> Epoch Te Tsae, B.C. 69–66: 1st year, 1st moon, 69, January.   *M. T. L.*

45   B.C. 61. *July.*

In the 1st year of the epoch Shin Tseo, the 6th moon, there was a comet in the east.

> Epoch Shin Tseo, B.C. 61–58: 1st year, 6th moon, 61, July.   *She Ke.*

46   B.C. 49. *April.*

In the 1st year of the epoch Han Lung, the 3rd moon, a strange star appeared to the north-east of Wang Leang : it was about 9 cubits in length. Its direction was to the west. It appeared between Ko Taou and Tsze Kung, into which it entered.

> Epoch Han Lung, B.C. 49 : 1st year, 3rd moon, 49, April.
> Wang Leang, β Cassiopeiæ.
> Ko Taou, ν, ξ, and others, in Cassiopeia.
> Tsze Kung, circle of perpetual apparition.   *She Ke.*

47   B.C. 48. *April.*

In the reign of Yuen Te, the 1st year of the epoch Choo Yuen, the 3rd moon, a strange star, resembling a large melon, was seen. Its colour was a bluish white. Its place was in Nan Tow, near the second star. It was about 4 cubits in length.

> Emperor Yuen Te, B.C. 48–33 ; epoch Choo Yuen, 48–44 : 1st year, 3rd moon, 48, April.
> Nan Tow, same as S. D. Tow, determined by ζ, τ, σ, and others, in Sagittarius.
>   *She Ke.*

48   B.C. 47. *June.*

In the 2nd year of the epoch Choo Yuen, the 5th moon, a comet was seen in the degrees of S. D. Maou. It was about 5 cubits to the east of Keuen She. Its colour was a bluish white. It was bright, and about $\frac{3}{10}$ths of a cubit in length.

>   Epoch Choo Yuen, B.C. 47: 2nd year, 5th moon, June.
>   S. D. Maou, determined by the Pleiades.
>   Keuen She, ε, ν, and others in Perseus.   *She Ke.*

49   B.C. 44.

In the 5th year of the same epoch a comet appeared in the north-east: its colour was a reddish yellow. It was 8 cubits in length. A few days after, its length was about 10 cubits. It was then in the north-east, pointing towards the S. D. Tsan. After about two months (?) it turned again to the west.

>   Epoch Choo Yuen, 5th year. B.C. 44.
>   S. D. Tsan, determined by α, β, and others in Orion.   *She Ke, M. T. L.*

>   The duration of this comet is doubtful.

50   B.C. 32. *February.*

In the reign of Ching Te, the 1st year of the epoch Keen Che, the 1st moon, there was a comet in Ying Shih: its colour was a bluish white. It was from 60 to 70 cubits in length, and about 1 cubit in width.

>   Emperor Ching Te, B.C. 32–47; epoch Keen Che, 32–29: 1st year, 1st moon, 32, February.
>   Ying Shih, same as S. D. Shih, determined by α Pegasi and others.   *M.T. L.*

51   B.C. 12. *August 26.*

In the 1st year of the epoch Yuen Yen, the 7th moon, day Sin Wei, there was a comet in the eastern part of S. D. Tsing. Its course was towards Woo Choo How. It appeared to the north of Ho Shoo, and advanced towards Heen Yuen and Tae Wei. It afterwards progressed at the rate of about 6 degrees in a day. In the morning it was seen in the east. On the 13th day, in the evening, it appeared in the west. It passed over the Tsze Fe and other neighbouring stars. It afterwards went into Ta Ho Tang, in the middle of Tsze Kung. It then passed round Teen Ho, and having left the boundaries of —— How it went to the south, and passed over Ta Keo and Che Te. It entered Teen She, and remained there during that lunation. It advanced slowly to the middle of Teen She, and afterwards left it to the west. On the 56th day it set with Tsang Lung.

   It is greatly to be regretted, that in the original work from which this account is taken many parts of the text are so indistinct, on account of injury to the block, that not only are some of the characters entirely obliterated, but others are so imperfect as to render their translation very uncertain, as they are almost illegible.

I have done the best I could under these circumstances, and believe the translation to be substantially correct. The notice of this comet in the 'She Ke' is exceedingly brief.

Epoch Yuen Yen, B.C. 12–9 : 1st year, B.C. 12, 7th moon : August. Day Sin Wei, August 26.

S. D. Tsin determined by γ, ε, λ, μ, &c. Geminorum.

Woo Choo How, θ, ν, τ, &c. Geminorum.

Ho Choo appears to be the same as Pih Ho, α, β, &c. Geminorum.

Heen Yuen, α, γ, η, and others in Leo and Leo Minor.

Tsze Fe, ζ, μ, ε Leonis.

Other characters, possibly names of stars, occur here which are not to be found in any of the lists I have seen : they, therefore, have not been identified.

Ho Tang not identified.

Tze Kung, circle of perpetual apparition.

—— How not identified, the preceding characters being illegible.

Ta Keo, Arcturus.   She Te, stars in the feet of Boötes.

Teen Ho, the Milky Way.

Tae Wei, space between stars in Leo and Virgo.

Teen She, space within the stars in Serpens.

Tsang Lung, the Azure Dragon ; one of the four divisions of the heavens, comprising our signs Libra, Scorpio, and Sagittarius.   *M. T. L.*

52   B.C. 5.   *March 5.*

In the reign of the Emperor Gae Te, the 2nd year of the epoch Keen Ping, the 2nd moon, a comet appeared in Keen New for about 70 days.

Emperor Gae Te, B.C. 6–1 ; epoch Keen Ping, 6–3 : 2nd year, 2nd moon, B.C. 5, March.

Keen New, same as S. D. New, determined by α, β, &c. Capricorni.   *M. T. L.*

53   B.C. 4.   *April.*

In the 3rd year of the same epoch, the 3rd moon, there was a comet in Ho Koo.

3rd year of epoch Keen Ping, B.C. 4 : 3rd moon, April.

Ho Koo, α, β, γ, &c. Aquilæ.   *Tung Keen.*

54   A.D. 13.   *December.*

In the reign of Wang Mang, the 5th year of the epoch Keen Kwo, the 11th moon, a comet appeared.

Wang Mang, a chieftain who usurped the Imperial dignity A.D. 9–22.

Epoch Keen Kwo, A.D. 9–13 : 5th year, 11th moon, A.D. 13, December.

                    She Ke.

55

In the 3rd year of the epoch Te Hwang, the 11th moon, there was a comet in S. D. Chang. It went to the south-east. After 5 days it was no longer seen.

Epoch Te Hwang, A.D. 20–22 : 3rd year, A.D. 22 : 11th moon, November.

S. D. Chang determined by κ, λ, μ Hydræ. *She Ke, M. T. L.*

56 A.D. 39. *March* 13.

In the reign of the Emperor Kwang Woo, the 15th year of the epoch Keen Woo, the 1st moon, on the day Ting Wei, a comet was seen in S. D. Maou. It was bright, 30 cubits in length, broad, and spreading like a tree. It went gradually to the north-west. It entered Ying Shih and passed into Le Kung. In the 2nd moon, on the day Yih Wei, it passed into the eastern part of S. D. Peih and disappeared. It was visible for 49 days.

Emperor Kwang Woo, A.D. 25–57 ; epoch Keen Woo, 25–55, 15th year, A.D. 39 : 1st moon, day Ting Wei, March 13 ; 2nd moon, day Yih Wei, April 30.

S. D. Maou determined by the Pleiades.

S. D. Peih determined by γ Pegasi and α Andromedæ.

S. D. Shih determined by α, β Pegasi, &c. Ying Shih, α Pegasi.

Le Kung, three groups of stars, of two each, in Pegasus, being λ μ, η ο, and ν, τ, and forming part of S. D. Shih. *She Ke, M. T. L.*

57 A.D. 55. *June* 4.

In the 30th year of the same epoch, in the intercalary moon, on the day Ke Woo, the planet Mercury being about 20 degrees in the eastern part of the S. D. Tsing, a white vapour appeared, pointing to the south-east. It was bright, and 10 cubits in length. It proved to be a comet. It went to the north-east. It passed above the western boundary of Tsze Kung. In the 5th moon, day Kea Tsze, it was no longer visible. It was seen altogether for 31 days.

Epoch Keen Woo, A.D. 55 : 30th year, intercalary moon. 'M. T. L.' informs us that this was the intercalary 4th moon, consequently the day Ke Woo is June 4.

5th moon, day Kea Tsze, July 4.

Tsze Kung, same as Tsze Wei Yuen, circle of perpetual apparition.

She Ke, M. T. L.

58 A.D. 60. *August* 9.

In the reign of Ming Te, the 3rd year of the epoch Yung Ping, the 6th moon, on the day Ting Maou, a comet appeared to the north of Teen Chuen. It was 2 cubits in length. It gradually went to the north, and entered the S. D. Kang to the south. It was visible 185 days.

Emperor Ming Te and epoch Yung Ping, A.D. 58–75 : 3rd year, A.D. 60, 6th moon, day Ting Maou, August 9.

S. D. Kang determined by ι, κ, λ, θ Virginis.

Teen Chuen, a, γ Persei, &c.   *She Ke, M. T. L.*

59   A.D. 61. *September* 27.

In the 4th year of the epoch Yung Ping, the 8th moon, day Sin Yew, a strange star appeared to the north-west of Kang Ho. It pointed towards Kwan Soo. It was visible for 70 days.

Epoch Yung Ping, A.D. 61 : 4th year, 8th moon, day Sin Yew, September 27.

Kang Ho, δ Boötis.   Kwan Soo, Corona Borealis.   *She Ke.*

60   A.D. 65.   *June* 4.

In the 8th year of the same epoch, the 6th moon, on the day Jin Woo, a comet appeared in the 37th degree of the S. D. Lew and Chang. It entered Heen Yuen and passed through Teen Chuen. It passed into Tae Wei. The vapour (tail) extended to Shang Keae. It was seen altogether for 56 days.

Yung Ping, 8th year, A.D. 65 : 6th moon, day Sin Woo, June 4.

S. D. Lew determined by δ, ε, and others in Hydra.

S. D. Chang determined by κ, λ, μ, and others in Hydra.

Heen Yuen, a, γ, ε, η, and others in Leo and Leo Minor.

Teen Chuen, a, γ, δ, and others in Perseus.

Shang Keae, possibly stars in Virgo.

Tae Wei, space between stars in Leo and Virgo.   *She Ke, M. T. L.*

61   A.D. 66. *February* 20.

In the 9th year of the same epoch, the 1st moon, day Woo Shin, a strange star appeared in S. D. New. It was 8 cubits in length. It passed through Keen Sing. It arrived at the south of S. D. Fang and then disappeared. It was visible 50 days.

Epoch Yung Ping, A.D. 66 : 9th year, 1st moon, day Woo Shin, Feb. 20.

S. D. New determined by a, β, &c. Capricorni.

S. D. Fang determined by β, δ, π, and others in Scorpio.

Keen Sing, ν, ξ, o, and others in Sagittarius.   *She Ke.*

62   A.D. 71. *March* 6.

In the 14th year of the same epoch, the 1st moon, day Woo Tsze, a strange star was seen for 60 days. It appeared first in S. D. Maou. It went into Heen Yuen and disappeared to the right of S. D. Keo.

Yung Ping, 14th year, A.D. 71 : 1st moon, day Woo Tsze, March 6.

S. D. Maou determined by the Pleiades.

S. D. Keo determined by a and ζ Virginis.

Heen Yuen, a, γ, ε, η, and others in Leo and Leo Minor.   *She Ke.*

63   A.D. 75. *July* 14.

In the 18th year of the epoch Yung Ping, the 6th moon, day Ke Wei, a comet appeared in S. D. Chang. It was 3 cubits in length. It turned and entered Lang Tseang. It passed into the south of Tae Wei.

> Yung Ping, A.D. 75 : 18th year, 6th moon, day Ke Wei, July 14.
> S. D. Chang determined by *a*, λ, μ, and others in Hydra.
> Tae Wei, space between stars in Leo and Virgo.
> Lang Tseang, Coma Berenices.   *She Ke, M. T. L.*

64   A.D. 76. *August* 9.

In the reign of Chang Te, the 1st year of the epoch Keen Choo, the 8th moon, day Kang Yin, a comet appeared in Teen She. It was 3 cubits in length. It passed on slowly into 3 degrees of Keen New. After 40 days it gradually disappeared.

> Emperor Chang Te, A.D. 76–88 ; epoch Keen Choo, 76–83, 1st year, 76 : 8th moon, day Kang Yin, August 9.
> Keen New for S. D. New, determined by *a*, β, and others in Capricornus.
> Teen She, space bounded by Serpens.   *She Ke, M. T. L.*

65   A.D. 77. *January* 23.

In the same year, the 12th moon, day Woo Yin, a comet appeared in 3 degrees of the S. D. Lew. Its length was from 8 to 9 cubits. It slowly entered Tsze Kung as far as the middle. After 106 days it gradually disappeared.

> Keen Choo, 1st year, A.D. 76 : 12th moon, day Woo Yin, A.D. 77, January 23.
> S. D. Lew determined by *a*, β, γ Arietis.
> Tsze Kung, circle of perpetual apparition.   *She Ke, M. T. L.*

> The 'She Ke' has the 11th moon of the 2nd year.

66   A.D. 84. *May* 25.

In the 1st year of the epoch Yuen Ho, the 4th moon, day Ting Sze, an extraordinary star appeared in the morning to the east. Its place was in the 18th degree of the S. D. Wei. It was 3 cubits in length. It passed over Ko Taou and entered Tsze Kung. On the 40th day it disappeared.

> Epoch Yuen Ho, A.D. 84–86 : 1st year, 84, 4th moon, day Ting Sze, May 25.
> S. D. Wei determined by the three stars in Musca.
> Ko Taou, *v*, ξ, *o*, π Cassiopeiæ.
> Tsze Kung, circle of perpetual apparition.   *She Ke.*

67   A.D. 102. *January* 7.

In the reign of the Emperor Ho Te, the 12th year of the epoch Yung Yuen, the 11th moon, on the day Kwei Yew, in the evening, a greenish-white vapour was seen,

E

about 30 cubits in length, commencing in Teen Yuen, to the north-east. It pointed to Keun She. It was seen altogether for 10 days.

Emperor Ho Te, A.D. 59–105; epoch Yung Yuen, 89–101 : 12th year, 101, 11th moon, day Kwei Yew, 102, January 7.

Teen Yuen, ι, κ, χ, φ Eridani.

Keun She, β Canis Majoris.   *She Ke.*

68   A.D. 110. *January.*

In the reign of the **Emperor Gan Te,** the 3rd year of the epoch Yung Choo, the 12th moon, a comet was seen to the south of Teen Yuen. It pointed towards the north-east. It was 6 or 7 cubits in length, and was of a greenish-white colour.

Emperor Gan Te, A.D. 107–125; epoch Yung Choo, 107–113 : 3rd year, 109, 12th moon, January.

Teen Yuen, γ, δ, ε, and others in Eridanus.

The Teen Yuen here mentioned must not be confounded with that in the preceding account, the characters being quite different, although of the same sound.

M. T. L.

69   A.D. 131.

In the reign of the Emperor Shun Te, the 6th year of the epoch Yung Keen, a comet appeared in S. D. Tow and Keen New. It disappeared in S. D. Heu and Wei.

Emperor Shun Te, A.D. 126–144; epoch Yung Keen, 126–131 : 6th year, 131.

S. D. Tow determined by ζ, τ, σ, and others in Sagittarius.

Keen New same as S. D. New, determined by α and others in Capricornus.

Wei determined by α Aquarii and γ Pegasi.   *M. T. L.*

Biot's date is 132.

70   A.D. 141. *March 27.*

In the 6th year of the epoch Yung Ho, the 2nd moon, day Ting Sze, a comet was seen in the east. It was 6 or 7 cubits in length. Its colour was a bluish white. It pointed south-west to Ying Shih, and extended to the stars Fun Moo. On the day Ting Chow the comet was about 1 degree in the S. D. Kwei. Its length was 6 cubits. On the day Kwei Hae it was seen in the morning to the north-west. It passed through the S. D. Maou and Peih. On the day Kea Shin it entered the eastern part of S. D. Tsing. It went on and passed through S. D. Kwei and Lew, and the seven stars in Chang. It was very bright, and extended to San Tae. It passed into the middle of Heen Yuen and then disappeared.   *She Ke, M. T. L.*

Epoch Yung Ho, A.D. 136–141 : 6th year, 141, 2nd moon, day Ting Sze, March 27. Other days : Ting Chow, April 16; Kwei Wei, April 22; Kea Shin, April 23.

S. D. Ying Shih. or Shih, determined by α, β Pegasi, &c.

Maou determined by the Pleiades.

S. D. Peih determined by a, γ, δ, &c. in Taurus.
S. D. Kwei determined by β, δ, ϵ, &c. in Andromeda and Pisces.
S. D. Lew determined by δ, ϵ, &c. Hydræ.
S. D. Chang determined by a, λ, μ, &c. Hydræ.
Fun Moo, γ, η, π Aquarii.
Heen Yuen, a and others in Leo and Leo Minor.

According to 'M. T. L.' this comet appeared in the epoch Yung Keen; the account being in other respects precisely the same. This would make the date 10 years earlier, viz. 131; and the days would be, Ting Sze, March 20; Ting Chow, April 9; Kwei Wei, April 15; and Kea Shin, April 16.

71   A.D. 149.   *October* 19.

In the reign of the Emperor Hwan Te, the 3rd year of the epoch Keen Ho, the 8th moon, day Yih Chow, a bright comet, 5 cubits in length, was seen in the middle of Teen She, to the south-east. Its colour was a yellowish white. In the 9th moon, on the day Woo Shin, it was no longer seen.

Emperor Hwan Te, A.D. 147–167; epoch Keen Ho, 147–149: 3rd year, 149, 8th moon, day Yih Chow, October 19; day Woo Shin, October 22.
Teen She, space bounded by Serpens.   *She Ke.*

'M. T. L.' has the 1st instead of the 3rd year. In the 'She Ke,' observations of Venus and other planets are recorded as having been made in the 1st and 2nd years of this epoch, and also in the 8th moon of the 3rd year. These are followed by the account of the comet as above. As the text in 'M. T. L.' is in other respects precisely similar, there is therefore no doubt as to the correctness of the year as given in the 'She Ke.'

72   A.D. 161.   *June* 14.

In the 4th year of the epoch Yen He, the 5th moon, on the day Sin Yew, a strange star was seen in Ying Shih. It progressed slowly. The tail became 5 cubits in length. It passed into the 1st degree of S. D. Sin. It turned and appeared as a comet.

Epoch Yen He, A.D. 158–166: 4th year, 161, 5th moon, day Sin Yew, June 14.
S. D. Ying Shih, or Shih, determined by a Pegasi, &c.
S. D. Sin, determined by a, σ, τ in Scorpio.   *She Ke, M. T. L.*

73   A.D. 178.   *September.*

In the reign of the Emperor Ling Te, the 1st year of the epoch Kwang Ho, the 8th moon, a comet appeared in the north of S. D. Kang. It passed into the middle of Teen She. It measured a cubit in length. It gradually increased in length until it measured from 50 to 60 cubits. Its colour was reddish. It passed through about 10 stellar divisions in about 80 days, and then disappeared in the middle of Teen Yuen.

Emperor Ling Te, A.D. 168–189; epoch Kwang Ho, 178–183: 1st year, 178, 8th moon, September.

S. D. Kang determined by ν, κ, λ, θ Virginis.
Teen She, space bounded by Serpens.
Teen Yuen, γ, δ, ε, and others in Eridanus.   *She Ke, M. T. L.*

74   A.D. 180.

In the 3rd year of the same epoch a comet appeared to the east of Lang Hoo. It entered into S. D. Chang and then disappeared.

Epoch Kwang Ho, 3rd year, A.D. 180.
S. D. Chang determined by κ, λ, μ, &c. Hydræ.
Lang Hoo, Sirius and other stars in Canis Major.   *She Ke, M. T. L.*

75   A.D. 182. *August.*

In the 5th year of the same epoch, the 7th moon, a comet appeared beneath San Tae. It went to the east. It entered Tae Wei. It passed Tae Tsze and Hing Chin. In about 20 days it disappeared.

Kwang Ho, 5th year, A.D. 182: 7th moon, August.
San Tae, the stars in the feet of Ursa Major.
Tae Wei, space between Virgo and Leo.
Tae Tsze, E Leonis.
Hing Chin, a star in Coma Berenices, near E Leonis.   *She Ke.*

'M. T. L.' has the 4th year of this epoch, A.D. 181.

76   A.D. 185. *December 7.*

In the 2nd year of the epoch Chung Ping, the 10th moon, on the day Kwei Hae, a strange star appeared in the middle of Nan Mun. It was like a large bamboo mat. It displayed the five colours, both pleasing and otherwise. It gradually lessened. In the 6th moon of the succeeding year it disappeared.

Epoch Chung Ping, A.D. 184–189: 2nd year, 185, 10th moon, day Kwei Hae, December 7, A.D. 186: 6th moon, July.
Nan Mun, α Centauri and stars near.   *She Ke.*

Biot's date is December 10, 173, and his epoch is Che Ping. In the 'She Ke' the epoch is precisely as here given, and no epoch Che Ping is to be found about this time in the Tables. The epoch in which the year 173 occurs is He Ping, but no comet or extraordinary star is recorded in the 'She Ke' as having appeared at that time. Biot's day would be quite correct for A.D. 173, but is not so for 185. No comet is to be found in 'M. T. L.' or the 'Tung Keen' under either of these dates.

77   A.D. 188. *March.*

In the 5th year of the same epoch, the 2nd moon, a comet appeared in S. D. Kwei. It went the contrary way and entered Tsze Kung. After having been seen for about 60 days it disappeared.

> Chung Ping, 5th year, A.D. 188 : 2nd moon, March.
> S. D. Kwei determined by β, δ, ε Andromedæ and stars in Pisces.
> Tsze Kung, the circle of perpetual apparition.   *She Ke.*

In 'M. T. L.' the account of this comet is placed under the epoch Kwang Ho, and Chung Ping does not appear as an epoch; but Kwang Ho occurs twice as an epoch, which is unusual. It appears, therefore, that there is a typographical error in 'M. T. L.,' and that Chung Ping should here be substituted for Kwang Ho. This would make the 'She Ke' and 'M. T. L.' consistent with each other.

78   A.D. 188. *July 29.*

In the 6th moon of the same year, day Ting Maou, a strange star, like a 3-shing measure, appeared in Kwan Soo. It went to the south-west. It entered Teen She, passed on to S. D. Wei, and disappeared.

> A.D. 188 : 6th moon, day Ting Maou, July 29.
> S. D. Wei determined by ε, μ, ν, and others in Scorpio.
> Teen She, space bounded by Serpens.
> Kwan Soo (also called Shih Soo), Corona Borealis.   *She Ke.*

A shing is described as a certain measure, containing 120,000 grains of millet.

Biot dates this comet, Kwang Ho, 5th year, 6th moon, 182, June 30. As in the former instance, the date I have given is that of the 'She Ke.' Three comets are recorded in the 'She Ke,' under the epoch Kwang Ho, which are not in Biot : they occur in the 1st (B.C. 178), the 3rd (B.C. 180), and 5th (B.C. 182) years of that epoch. In the 'Tung Keen Kang Muh' they are also given under the epoch Kwang Ho, as well as that of the 5th year of Chung Ping (B.C. 188, July); also not in Biot. That of the 2nd year, and the present one, do not occur in the 'Tung Keen' under the epoch Chung Ping. They are not in 'M. T. L.'

79   A.D. 192. *October.*

In the reign of the Emperor Heen Te, the 3rd year of the epoch Choo Ping, the 9th moon, a comet was seen. It was about 100 cubits in length. Its colour was white. It appeared to the south of the S. D. Keo and Kang.

> Emperor Heen Te, A.D. 190–220; epoch Choo Ping, 190–193: 9th moon, October.
> S. D. Keo determined by α Virginis and another.
> S. D. Kang determined by ι, κ, λ, μ, ρ Virginis.   *M. T. L.*

80   A.D. 193. *November.*

In the 4th year of the same epoch, the 10th moon, a comet appeared between the two Keos. It went to the north-east. It entered Teen She as far as the middle, and disappeared.

> Choo Ping, 4th year, 193: 10th moon, November.
> The two Keos. S. D. Keo determined by *a* Virginis and another. Ta Keo, Arcturus.
> Teen She, the space bounded by Serpens.   *She Ke, M. T. L.*

81   A.D. 200. *November* 7.

In the 5th year of the epoch Keen Gan, the 10th moon, day Sin Hae, there was a comet in Ta Leang.

> Epoch Keen Gan, A.D. 196–220: 5th year, 200, 10th moon, day Sin Hae, November 7th.
> Ta Leang, unascertained.   *M. T. L.*

82   A.D. 204. *December.*

In the 9th year of the same epoch, the 11th moon, a comet appeared in the eastern part of S. D. Tsing, near to S. D. Kwei. It entered Heen Yuen and Tae Wei.

> Epoch Keen Gan, 9th year, A.D. 204: 11th moon, December.
> S. D. Tsing determined by δ, ε, λ, &c. Geminorum.
> S. D. Kwei determined by γ, δ, η, θ Cancri.
> Heen Yuen, *a* and others in Leo and Leo Minor.
> Tae Wei, space in Leo and Virgo.   *She Ke, M. T. L.*

83   A.D. 206. *February.*

In the 11th year, 1st moon, there was a comet in Pih Tow. The head was in the middle of that asterism. It was also seen in S. D. Wei, in Kwan, and in Tsze Kung, and in the morning it extended towards the north.

> Keen Gan, 11th year, A.D. 206: 1st moon, February.
> S. D. Wei determined by ε, μ, ν, and others in Scorpio.
> Kwan, possibly Corona Borealis.
> Tsze Kung, circle of perpetual apparition.
> Pih Tow, the seven bright stars in Ursa Major.   *She Ke, M. T. L.*

84   A.D. 207. *November* 10.

In the 12th year, 10th moon, day Sin Maou, there was a comet in Shun Wei.

> Keen Gan, 12th year, A.D. 207: 10th moon, day Sin Maou, November 10.
> Shun Wei, one of the twelve kung, or signs, answering to Virgo.
>   *She Ke, M. T. L.*

85   A.D. 213. *January.*

In the 17th year, 12th moon, there was a comet in Woo Choo How.

    Keen Gan, 17th year, A.D. 213: 12th moon, January.
    Woo Choo How, θ, ι, ν, τ, ϕ Geminorum.   *She Ke, M. T. L.*

86   A.D. 218. *April.*

In the 23rd year of the same epoch, in the 3rd moon, a comet was seen in the morning, in the east, for about 20 days. In the evening it appeared in the west. It entered and passed through Woo Chay, Tung Tsing, Woo Choo How, Wan Chang, Heen Yuen, How Fe, and Tae Wei. It was pointed and bright. Its course was towards Te Tso.

    Keen Gan, 23rd year, A.D. 218: 3rd moon, April.
    Woo Chay, a, β, o, κ Aurigæ, and β Tauri.
    Tung Tsing, the eastern part of S. D. Tsing, determined by δ, ε, λ, &c. Geminorum.
    Woo Choo How, θ, ι, ν, τ, ϕ Geminorum.
    Wan Chang, θ, ϕ, ν Ursæ Majoris.
    Heen Yuen, a and others in Leo and Leo Minor.
    How Fe, unascertained.
    Te Tso, a Herculis.
    Tae Wei, space between stars in Leo and Virgo.   *She Ke, M. T. L.*

87   A.D. 236. *November.*

In the reign of How Choo, the 14th year of the epoch Keen Hing, there was a comet in the east.

    How Choo, 223–264; epoch Keen Hing, 223–237: 14th year, 236, 10th moon, November.

    For further particulars respecting this comet see No. 91.

At the close of the Han dynasty China was divided into three principal states, Wei, Woo, and Shuh. This was the celebrated period of the San Kwo, or Three Nations. The Shuh was a branch of the Han, and under the name of the How, or later Han, has a place among the regular dynasties. It maintained the supreme power until A.D. 264, when the Wei, until then a minor state, obtained the superiority, and founded a new dynasty, under the name of the Tsin. The comets which immediately follow are those observed during the Wei, A.D. 220–264, and the Tsin, 265–419. These are succeeded by the comets observed during the Sung, Tze, Leang, Chin, and Suy dynasties, embracing the period between A.D. 420 and 617, when the Tang dynasty obtained the superiority.

WEI, A MINOR STATE, A.D. 220–264.

88 A.D. 222. *November 4.*

In the reign of Wan Te, the 3rd year of the epoch Hwang Choo, the 9th moon, day Kea Shin, a strange star was seen in Tae Wei, to the left, within Yih Mun.

>Wan Te, A.D. 220–226; epoch, Hwang Choo, 220–226: 3rd year, 222, 9th moon, day Kea Shin, November 4th.
>Tae Wei, space bounded by stars in Leo and Virgo.
>Yih Mun, space between η and β Virginis. *She Ke.*

89 A.D. 225. *December 9.*

In the 6th year of the same epoch, the 10th moon, on the day Yih Wei, there was a comet in Shaou Wei. It passed through Heen Yuen.

>Epoch Hwang Choo, 6th year, A.D. 225: 10th moon, day Yih Wei, Dec. 9.
>Shaou Wei, same as Tae Wei, space between stars in Leo and Virgo.
>Heen Yuen, α and others in Leo and Leo Minor. *M. T. L.*

90 A.D. 232. *December 4.*

In the reign of Ming Te, the 6th year of the epoch Tae Ho, the 11th moon, day Ping Yin, there was a comet in S. D. Yih, near the star Shang Tseang in Tae Wei.

>Ming Te, A.D. 227–239; epoch Tae Ho, 227–232: 6th year, 232, 11th moon, day Ping Yin, December 4th.
>S. D. Yih determined by α and others in Crater.
>Tae Wei, space between stars in Leo and Virgo.
>Shang Tseang, γ Virginis. *She Ke, M. T. L.*

91 A. D. 236. *November 30.*

In the 4th year of the epoch Tsing Lung, the 10th moon, on the day Kea Shin, there was a comet in Ta Shin. It was 3 cubits in length. On the day Yih Yew the comet was in the east. In the 11th moon, day Yih Hae, the comet was seen passing near the stars Hwan Chay and Teen Ke.

>Epoch Tsing Lung, A.D. 233–236: 4th year, 236, 10th moon, day Kea Shin, November 30th; days Yih Yew, December 1st; Yih Hae, 237, January 20th.
>Ta Shin. The Commentary intimates that Ta Shin is the same as Teen Wang—Polaris.
>Hwan Chay, small stars in head of Ophiuchus. *She Ke, M. T. L.*
>Teen Ke, small stars near θ Herculis.
>The 'She Ke' has Ke Hae for Yih Hae, which would be 236, December 15.
>This appears to be the same comet as No. 87.

92   A.D. 238. *September.*

In the 2nd year of the epoch King Choo, the 8th moon, a comet was seen in S. D. Chang. It was 3 cubits in length. It went backwards towards the west. On the 41st day it disappeared.

        Epoch King Choo, A.D. 237–239 : 2nd year, 238, 8th moon, September.
        S. D. Chang determined by κ, λ, μ, &c. Hydræ.   *She Ke, M. T. L.*

93   A.D. 238. *November 29.*

In the 10th moon of the same year, on the day Kwei Sze, a strange star was seen in S. D. Wei. It went the contrary way. Its place was to the north of Le Kung, and to the south of Tang Shay. On the day Kea Shin it was near the star Tsung; on the day Ke Yew it disappeared.

        King Choo, 2nd year, A.D. 238 : 10th moon, days, Kwei Sze, November 29th ; Kea Shin, December 10th; Ke Yew, December 15th.
        S. D. Wei determined by α Aquarii and θ, ε Pegasi.
        Le Kung, three groups of two stars, each in Pegasus, with α and β Pegasi they form S. D. Shih.
        Tang Shay, stars in Cygnus, Lacerta, and Andromeda.
        Tsung Ting, small stars in head of Taurus Poniatowski.   *She Ke.*

94   A.D. 240. *November 5.*

In the reign of Fei Te, in the 1st year of the epoch Ching Che, day Yih Yew, a comet was seen in the west. Its place was in S. D. Wei. It was 20 cubits in length. It passed through S. D. New, near to Tae Pih. In the 11th moon, day Kea Tsze, it entered Yu Lin.

        Emperor Fei Te, A.D. 240–253 ; epoch Ching Che, 240–248, 1st year, A.D. 240, 10th moon, day Yih Yew, Nov. 5th ; 11th moon, day Kea Tsze, Dec. 14th.
        S. D. Wei determined by ε, μ, ν, &c. in Scorpio.
        S. D. New determined by ζ, τ, ς, &c. in Sagittarius.
        Tae Pih, the planet Venus.
        Yu Lin, δ, τ, χ, and others in Aquarius and Pisces.   *She Ke, M. T. L.*

95   A.D. 245. *September* 18.

In the 6th year of the same epoch, the 8th moon, day Woo Woo, a comet was seen among the seven stars of S. D. Sing. It was 2 cubits in length. Its colour was white. It passed into the S. D. Chang. After 23 days it disappeared.

        Ching Che, 6th year, A.D. 245 : 8th moon, day Woo Woo, September 18th.
        S. D. Sing determined by the seven stars in α Hydræ and others near.
        S. D. Chang determined by κ, λ, μ, &c. in Hydra.   *She Ke, M. T. L.*

96   A.D. 247. *January* 16.

In the 7th year of the same epoch, the 11th moon, on the day Kwei Hae, a comet was seen in S. D. Chin. It was 1 cubit in length. It was visible for 156 days, and then disappeared.

>Ching Che, 7th year, A.D. 246: 11th moon, day Kwei Hae, 247, Jan. 16th.
>S. D. Chin determined by β and others in Corvus.   *She Ke, M. T. L.*

'M. T. L.' has 56 instead of 156 days, during which the comet was seen; in which he is followed by Biot, and which appears to be the more probable number. The 'She Ke' is as above.

97   A.D. 248. *April.*

In the 9th year of the same epoch, the 3rd moon, there was (a comet) seen in S. D. Maou. It was 6 cubits in length: its colour was a bluish white. The tail pointed to the south-west. In the 7th moon it was seen in S. D. Yih, and was 2 cubits in length. It passed into S. D. Chin: after 42 days it disappeared.

>Ching Che, 9th year, A.D. 248: 3rd moon, April; 7th moon, August.
>S. D. Maou determined by the Pleiades.
> Yih determined by a, &c. Crateris.
> Chin determined by β, &c. Corvi.   *She Ke, M. T. L.*

98   A.D. 251. *December* 21.

In the 3rd year of the epoch Kea Ping, the 11th moon, day Kwei Hae, there was a comet in Ying Shih. It went to the west, and was visible for 90 days, when it disappeared.

>Epoch Kea Ping, A.D. 249–253: 3rd year, 251, 11th moon, day Kwei Hae, December 21st.
>S. D. Ying Shih, same as Shih, determined by a, β Pegasi, &c.
>   *She Ke, M. T. L.*

99   A.D. 252. *March* 25.

In the 4th year of the same epoch, the 2nd moon, day Ting Yew, a comet was seen in the west. Its place was in S. D. Wei. It was from 50 to 60 cubits in length: its colour white. The tail pointed to the south, passing through S. D. Tsan. It was visible for 20 days, and then disappeared.

>Epoch Kea Ping, 4th year, 252: 2nd moon, day Ting Yew, March 25th.
>S. D. Wei determined by the three stars in Musca.
>S. D. Tsan determined by a, β, &c. Orionis.   *She Ke, M. T. L.*

100   A.D. 253. *December.*

In the 5th year of the same epoch, the 11th moon, a comet was seen in S. D. Chin. It was 50 cubits in length. Its place was in Tae Wei, to the left of Tso Che Fa. It pointed towards the south-east. It was visible for 190 days, when it disappeared.

Kea Ping, 253: 5th year, 11th moon, December.
S. D. Chin determined by β, &c. Corvi.
Tae Wei, space between stars in Leo and Virgo.
Tso Che Fa, η Virginis.   *She Ke, M. T. L.*

101   A.D. 255. *February.*

In the reign of Shaou Te, the 2nd year of the epoch Ching Yuen, the 1st moon, there was a comet in Woo Yue, to the north-west, in the horizon.

Emperor Shaou Te, A.D. 254–259; epoch Ching Yuen, 254–255: 2nd year, 1st moon, 255, February.

Woo Yue, ε, ζ Aquilæ.   *She Ke, M. T. L.*

102   A.D. 257. *December.*

In the 2nd year of the epoch Kan Loo, the 11th moon, a comet was seen in S. D. Keo. Its colour was white.

Epoch Kan Loo, A.D. 256–259: 2nd year, 11th moon, December.
S. D. Keo determined by α and another in Virgo.   *She Ke, M. T. L.*

103   A.D. 259. *November 23.*

In the 4th year of the same epoch, 10th moon, day Ting Chow, a strange star was seen in the middle of Tae Wei. It turned and went to the south-east. It passed through S. D. Chin. It was altogether visible for 7 days, and then disappeared.

Kan Loo, 4th year, A.D. 259: 10th moon, day Ting Chow, November 23rd.
Tae Wei, space in Leo and Virgo.
S. D. Chin determined by β and others in Corvus.   *She Ke.*

104   A.D. 262. *December 2.*

In the reign of Yuen Te, the 3rd year of the epoch King Yuen, 11th moon, day Jin Yin, a comet was seen in S. D. Kang. Its colour was white. It was $\frac{5}{10}$ths of a cubit in length. It went to the north. After 45 days it disappeared.

Emperor Yuen Te, A.D. 260–265; epoch King Yuen, 260–263: 3rd year, 262, 11th moon, day Jin Yin, December 2nd.
S. D. Kang determined by ι, κ, λ, θ Virginis.   *She Ke, M. T. L.*

105   A.D. 265. *June.*

In the 2nd year of the epoch Han He, the 5th moon, a comet was seen in Wang Leang. Its length was about 10 cubits. Its colour was white. It pointed towards the south-east. After 12 days it disappeared.

Epoch Han He, A.D. 264–265: 2nd year, 5th moon, 265, June.
Wang Leang, α, β, η, κ Cassiopeiæ.   *She Ke, M. T. L.*

The WEI having obtained the superiority adopted the name of TSIN, and founded the

TSIN DYNASTY, A.D. 265-419.

106   A.D. 268. *February* 18.

In the reign of the Emperor Woo Te, the 4th year of the epoch Tae Che, the 1st moon, day Ping Seuh, a comet was seen in S. D. Chin. Its colour was a bluish white. It went to the north-west, but afterwards turned and went to the east.

> Emperor Woo Te, A.D. 265–289; epoch Tae Che, 265–274: 4th year, 268, 1st moon, day Ping Seuh, February 18th.
> S. D. Chin determined by β and others in Corvus.   *She Ke, M. T. L.*

107   A.D. 275. *January.*

In the 10th year of the same epoch, the 12th moon, there was a comet in S. D. Chin.
> Epoch Tae Che, 10th year, 274: 12th moon, 275, January.
> S. D. Chin determined by β and others in Corvus.

108   A.D. 276. *June* 24.

In the 2nd year of the epoch Han Ning, the 6th moon, on the day Kea Seuh, there was a comet in S. D. Te. In the 7th moon the comet was near Ta Keo. In the 8th moon the comet was in Tae Wei. It passed into S. D. Yih, and also into Pih Tow and San Tae.

> Epoch Han Ning, A.D. 275–279: 2nd year, 276, 6th moon, day Kea Seuh, June 24; 7th moon, July; 8th moon, August.
> S. D. Te determined by a, β, γ, &c. Libræ.
> S. D. Yih determined by a and others in Crater.
> Ta Keo, Arcturus.   Pih Tow, the seven bright stars in Ursa Major.
> San Tae, the stars in the feet of Ursa Major.   *She Ke, M. T. L.*

In the original this account is divided into three parts, separated by astrological inferences. There is no doubt but that they relate to one comet. The S. D. also are here considered as extending to the Pole. The same remarks apply to the next two comets.

109   A.D. 277. *February.*

In the 3rd year of the same epoch, the 1st moon, there was a comet in the west. In the 3rd moon it was in S. D. Wei. In the 4th moon the comet was in Neu Yu. In the 5th moon it was in the east. In the 7th moon it was in Tsze Kung.

> Epoch Han Ning, 3rd year, A.D. 277: 1st moon, February; 3rd moon, April; 4th moon, May; 5th moon, June; 7th moon, August.
> S. D. Wei determined by the three stars in Musca.
> Neu Yu, π Leonis.
> Tsze Kung, circle of perpetual apparition.   *She Ke, M. T. L.*

110   A.D. 279. *April.*

In the 5th year of the same epoch, the 3rd moon, there was a comet in S. D. Lew. In the 4th moon the comet was in New Yu.   In the 7th moon the comet was in Tsze Kung.

> Han Ning, 5th year, A.D. 279 : 3rd moon, April ; 4th moon, May ; 7th moon, August.
> S. D. Lew determined by δ, ε, &c. Hydræ.   New Yu, π Leonis.
> Tsze Kung, circle of perpetual apparition.   *She Ke, M. T. L.*

> There is evidently here some confusion in the original text, as the observations of the 4th and 7th moons are precisely the same as the observations of the preceding comet in the 4th and 7th moons of its appearance.   It is, however, the same both in the 'She Ke' and 'M. T. L.'

111   A.D. 281. *September.*

In the 2nd year of the epoch Tae Kung, the 8th moon, there was a comet in S. D. Chang.

> Epoch Tae Kung, A.D. 280–289 : 2nd year, 281, 8th moon, September.
> S. D. Chang determined by κ, λ, μ, &c. Hydræ.   *She Ke, M. T. L.*

112   A.D. 281. *December.*

In the 11th moon of the same year there was a comet in Heen Yuen.

> A.D. 281, 11th moon, December.
> Heen Yuen, α and other stars in Leo and Leo Minor.   *She Ke, M. T. L.*

113   A.D. 283. *April 22.*

In the 4th year of the same epoch, the 3rd moon, day Woo Shin, there was a comet in the south-west.

> Tae Kung, 4th year, A.D. 283 : 3rd moon, day Woo Shin, April 22.
>   *She Ke, M. T. L.*

114   A.D. 287. *September.*

In the 8th year of the same epoch, the 9th moon, there was a comet in Nan Tow. Its length was reckoned at 100 cubits.   In about 10 days it disappeared.

> Tae Kung, 8th year, A.D. 287 : 9th moon, September.
> Nan Tow, same as S. D. Tow, determined by ζ, τ, σ, &c. Sagittarii.
>   *She Ke, M. T. L.*

H

115   A.D. 290. *May.*

In the 1st year of the epoch Tae He, the 4th moon, there was a strange star in Tsze Kung.

> Epoch Tae He, A.D. 290. In the Tables this epoch is written Yung He, and is made the 1st of the Emperor Hwuy Te: 4th moon, May.
>
> Tsze Kung, circle of perpetual apparition.   *She Ke, M. T. L.*

116   A.D. 296. *May.*

In the reign of Hwuy Te, the 5th year of the epoch Yuen Kang, the 4th moon, there was a comet in S. D. Kwei. It passed into Heen Yuen and Tae Wei. It crossed San Tae and Ta Ling.

> Emperor Hwuy Te, A.D. 290–306; epoch Yuen Kang, 291–299: 5th year, 296: 4th moon, May.
>
> S. D. Kwei determined by β, δ, ε, &c. Andromedæ and stars in Pisces.
>
> Heen Yuen, α and others in Leo and Leo Minor.
>
> Tae Wei, space between stars in Leo and Virgo.
>
> San Tae, stars in feet of Ursa Major. Ta Ling, γ and others in Perseus.
>
>   *She Ke, M. T. L.*

117   A.D. 300. *April.*

In the 1st year of the epoch Yung Kang, 3rd moon, a strange star was seen in the south.

> Epoch Yung Kang, A.D. 300: 3rd moon, April. *She Ke.*
>
> Possibly a meteor.

118   A.D. 301. *January.*

In the 12th moon of the same year a comet appeared in S. D. New, to the west. It pointed to Tien She.

> A.D. 300, 12th moon, 301, January.
>
> S. D. New determined by α, β, &c. Capricorni. *She Ke, M. T. L.*

119   A.D. 301. *May.*

In the 2nd year of the same epoch, 4th moon, a comet was seen near the star Tse.

> Yung Kang, 2nd year, 301: 4th moon, May.
>
> Star Tse, H Herculis.   *She Ke, M. T. L.*

120   A.D. 302. *May.*

In the 1st year of the epoch Tae Gan, the 4th moon, a comet was seen in the daytime.

> Tae Gan, 302–303, 1st year, 4th moon, 302, May. *She Ke, M. T. L.*

121   A.D. 303. *April.*

In the 2nd year of the same epoch, the 3rd moon, a comet was seen in the east. It pointed towards San Tae.

     Tae Gan, 2nd year, A.D. 303: 3rd moon, April.
     San Tae, the stars in feet of Ursa Major.   *She Ke, M. T. L.*

122   A.D. 304. *May.*

In the 1st year of the epoch Yung Hing, the 5th moon, there was a strange star in S. D. Peih.

     Epoch Yung Hing, A.D. 304–305: 1st year, 304: 5th moon, May.
     S. D. Peih determined by a, γ, δ, &c. Tauri.   *She Ke.*

123   A.D. 305. *September.*

In the 2nd year of the same epoch, the 8th moon, there was a comet in S. D. Maou and Peih.

     Yung Hing, 2nd year, 8th moon, A.D. 305, September.
     S. D. Maou determined by the Pleiades.
     S. D. Peih determined by a, γ, δ, ε, &c. Tauri.   *She Ke, M. T. L.*

124   A.D. 305. *November* 21.

In the 10th moon of the same year, on the day Ting Chow, there was a comet in Pih Tow, near the star Seuen Ke.

     A.D. 305, 10th moon, day Ting Chow, November 21st.
     Pih Tow, the seven bright stars in Ursa Major.
     Seuen Ke, same as Teen Ke, γ Ursæ Majoris.   *She Ke, M. T. L.*

125   A.D. 329. *August.*

In the reign of Ching Te, the 4th year of the epoch Han Ho, the 7th moon, there was a comet in the north-west. It entered into S. D. Tow. After 23 days it disappeared.

     Emperor Ching Te, A.D. 326–342; epoch Han Ho, 326–334: 4th year, 329: 7th moon, August.
     S. D. Tow determined by ζ, τ, σ, ϕ, &c. Sagittarii.   *She Ke, M. T. L.*

126   A.D. 336. *February* 16.

In the 2nd year of the epoch Han Kang, 1st moon, day Sin Sze, a comet was seen in the evening, in the west. Its place was in S. D. Kwei.

     Epoch Han Kang, A.D. 335–342: 2nd year, 336: 1st moon, day Sin Sze, Feb. 16.
     S. D. Kwei determined by β, δ, ε, &c. Andromedæ and others in Pisces.
                                           She Ke, M. T. L.

     'M. T. L.' has the 2nd moon, March; but no day Sin Sze occurs in March in that year.

127   A.D. *340. March* 5.

In the 2nd moon of the 6th year of the same epoch, day Kang Shin, there was a comet in Tae Wei.

> Epoch Han Kang, 6th year, 340 : 2nd moon, day Kang Shin, March 5th.
> Tae Wei, space between stars in Leo and Virgo.   *She Ke, M. T. L.*

128   A.D. *343. December* 8.

In the reign of the Emperor Kang Te, the 1st year of the epoch Keen Yuen, the 11th moon, 6th day, a comet was seen in S. D. Kang. Its length was 7 cubits. Its colour was white.

> Emperor Kang Te and epoch Keen Yuen, A.D. 343–344 : 1st year, 343 : 11th moon, 6th day, December 8th.
> S. D. Kang determined by ι, κ, λ, θ Virginis.   *She Ke, M. T. L.*

129   A.D. *349. November* 23.

In the reign of Muh Te, the 5th year of the epoch Yung Ho, the 11th moon, day Yih Maou, a comet was seen in S. D. Kang. It was bright, and directed towards the west. Its colour was white. It was 10 cubits in length. In the 1st moon of the 6th year, on the day Ting Chow, the comet was still visible in S. D. Kang.

> Emperor Muh Te, A.D. 345–361 ; epoch Yung Ho, 345–356 : 5th year, 349, 11th moon, day Yih Maou, 349, November 23rd; 6th year, 350 : 1st moon, day Ting Chow, February 13th.
> S. D. Kang determined by ι, κ, λ, θ Virginis.   *She Ke, M. T. L.*

130   A.D. *358. July* 1.

In the 2nd year of the epoch Shing Ping, the 5th moon, day Ting Hae, a comet was seen in Teen Chuen, in S. D. Wei.

> Epoch Shing Ping, A.D. 357–361 : 2nd year, 358 : 5th moon, day Ting Hae, July 1.
> Teen Chuen, γ, η Persei.
> S. D. Wei determined by the three stars in Musca.   *She Ke, M. T. L.*

131   A.D. *363. August.*

In the reign of Gae Te, the 1st year of the epoch Hing Ning, the 8th moon, there was a comet in S. D. Keo and Kang. It entered the boundary of Teen She.

> Emperor Gae Te, A.D. 362–365 ; epoch Hing Ning, 363–365 : 1st year, 363 : 8th moon, August.
> S. D. Keo determined by α Virginis and another.
> S. D. Kang determined by ι, κ, λ, θ Virginis.
> Teen She, space bounded by Serpens.   *She Ke, M. T. L.*

132   A.D. 369. *March.*

In the reign of Te Yih, the 4th year of the epoch Tae Ho, the 2nd moon, a strange star was seen in Tsze Kung, near its western boundary. In the 7th moon it disappeared.

Emperor Te Yih and epoch Tae Ho, A.D. 366–370: 4th year, 369: 2nd moon, March.

Tsze Kung, circle of perpetual apparition.   *She Ke.*

133   A.D. 373. *March 9.*

In the reign of the Emperor Haou Woo, the 1st year of the epoch Ning Kang, the 1st moon, day Ting Sze, there was a comet in S. D. Neu, Heu, Te, Kang, Keo, Chin, Yih, and Chang. In the 2nd moon, day Ping Seuh, the comet was seen in S. D. Te. In the 9th moon, day Ting Chow, the comet was in Teen She.

Emperor Haou Woo, A.D. 373–396; epoch Ning Kang, 373–375: 1st year, 373, 1st moon, day Ting Sze, March 9th; 2nd moon, day Ping Seuh, April 7th; 9th moon, day Ting Chow, September 25th.

S. D. Neu determined by ε, μ, &c. Aquarii.

Heu determined by β Aquarii and others.

Te determined by a, β, &c. Libræ.

Keo determined by a and ξ Virginis.

Kang determined by ι, κ, λ, θ Virginis.

Chin determined by β, &c. Corvi.

Yih determined by a and others in Crater.

Chang determined by κ, λ, μ Hydræ.

Teen She, the space bounded by Serpens.   *M. T. L.*

The 'She Ke' has this comet under the 2nd year, 1st and 3rd moons. This would make it A.D. 374, February and March.

134   A.D. 386. *April.*

In the 11th year of the epoch Tae Yuen, the 3rd moon, there was a comet in Nan Tow. It was visible until the 6th moon, when it disappeared.

Epoch Tae Yuen, A.D. 376–396: 11th year, 386, 3rd moon, April: 6th moon, July.

Nan Tow, same as S. D. Tow, determined by λ, μ, ϕ, σ, &c. Sagittarii.

                                                                She Ke.

135   A.D. 390. *August 22.*

In the 15th year of the same epoch, the 7th moon, day Jin Shin, there was a comet in Pih Ho. It crossed Tae Wei, San Tae, and Wan Chung. It entered Pih Tow. Its colour was white. It was about 100 cubits in length. In the 8th moon, on the day Woo Seuh, it entered Tsze Wei and disappeared.

I

Tae Yuen, 15th year, A.D. 390 : 7th moon, day Jin Shin, August 22nd ; day Woo Seuh, September 17th.

Pih Ho, *a*, *β*, &c. Geminorum.

Tae Wei, space between stars in Leo and Virgo.

San Tae, stars in feet of Ursa Major.

Wan Chang, *θ*, *v*, *φ* Ursæ Majoris.

Pih Tow, the seven bright stars in Ursa Major.

Tsze Wei, circle of perpetual apparition.   *She Ke, M. T. L.*

136   A.D. *393. March.*

In the 18th year of the same epoch, the 2nd moon, a strange star appeared in the middle of S. D. Sing.   In the 9th moon it disappeared.

Tae Yuen, 18th year, A.D. 393 : 2nd moon, March ; 9th moon, October.

S. D. Sing determined by *a*, *τ*, &c. Hydræ.   *She Ke.*

137   A.D. 400. *March 19.*

In the reign of Gan Te, the 4th year of the epoch Lung Gan, the 2nd moon, day Ke Chow, there was a comet in S. D. Kwei.   It was more than 30 cubits in length.   It was above Ko Taou, in the western part of Tsze Kung.   It entered Pih Tow Kwei. It passed on to San Tae.   In the 3rd moon it entered Tae Wei, Te Tso, and Twan Mun.

Emperor Gan Te, A.D. 397–418 ; epoch Lung Gan, 397–400 : 4th year, 400 : 2nd moon, day Ke Chow, March 19 ; 3rd moon, April.

S. D. Kwei determined by *β*, *δ*, *ε* Andromedæ and stars in Pisces.

Ko Taou, *δ*, *ε*, and others in Cassiopeia.

Tsze Kung, circle of perpetual apparition.

Pih Tow, the seven bright stars in Ursa Major.

San Tae, stars in the feet of Ursa Major.

Tae Wei, space between stars in Leo and Virgo.

Te Tso, or Woo Te Tso, *β* Leonis and stars near.

Twan Mun, possibly Teen Mun, between *β* and *η* Virginis.   *She Ke, M. T. L.*

138   A.D. 401. *January 2.*

In the 12th moon of the same year, on the day Woo Yin, there was a comet in Shih Soo, Teen She, and Teen Tsin.

A.D. 400 : 12th moon, day Woo Yin, 401, January 2nd.

Shih Soo, Corona Borealis.

Teen She, space bounded by Serpens.

Teen Tsin, *a*, *β*, *ε*, &c. Cygni.

139   A.D. 402. *November* 12.

In the 1st year of the epoch Yuen Hing, the 10th moon, a strange star appeared. Its colour was white. It resembled a handful of meal. Its place was to the west of Tae Wei. In the 12th moon it entered Tae Wei.

    Epoch Yuen Hing, A.D. 402–404, 1st year, 402 : 10th moon, November ; 12th moon, January, 403.

    Tae Wei, space between stars in Leo and Virgo.   *She Ke.*

140   A.D. 415. *June* 24.

In the 11th year of the epoch E He, the 5th moon, day Kea Shin, two comets appeared in Teen She. They swept Te Tso. They were in the north of S. D. Fang and Sin.

    Epoch E He, A.D. 405–418 : 11th year, 415: 5th moon, day Kea Shin, June 24.
    S. D. Fang determined by β, δ, &c. in Scorpio.
    Sin determined by a, σ, τ in Scorpio.
    Teen She, space bounded by Serpens.
    Te Tso, a Herculis.   *She Ke, M. T. L.*

141   A.D. 418. *September* 15.

In the 14th year of the same epoch, 5th moon, day Kang Tsze, there was a comet in Pih Tow Kwei, towards the middle. In the 7th moon, day Kwei Hae, the comet appeared in the western part of Tae Wei, above Juy Ke, and below the star Leang. It was bright, and gradually lengthened until it was about 100 cubits in length. In its course it swept Pih Tow, Tsze Wei, and Chung Tae.

    E He, 14th year, A.D. 418 : 7th moon, day Kwei Hae, September 15th.
    Tae Wei, space between stars in Leo and Virgo.
    Juy Ke unascertained.
    Seang. Several stars having this name occur in Tae Wei : one of these, to the west, is possibly that here referred to.
    Pih Tow, the seven bright stars in Ursa Major. Kwei in Pih Tow is referred to the square in the same.
    Chung Tae, λ, μ Ursæ Majoris.
    Tsze Wei, circle of perpetual apparition.   *She Ke, M. T. L.*

142   A.D. 419. *February* 7.

In the reign of Kung Te, the 1st year, 1st moon, day Woo Seuh, there was a comet in the western boundary of Tae Wei.

    Emperor Kung Te and 1st year, 419 : 1st moon, day Woo Seuh, Feb. 17.
    Tae Wei, space between stars in Leo and Virgo.   *She Ke, M. T. L.*

    Kung Te was the last Emperor of the Tsin dynasty : he reigned but one year, and was succeeded by the Sung.

THE EARLY SUNG DYNASTY, A.D. 420–478.

143   A.D. 422. *March* 21.

In the reign of Woo Te, the 3rd year of the epoch Yung Choo, the 2nd moon, day Ping Seuh, a comet was seen in S. D. Heu and Wei.

> Emperor Woo Te and epoch Yung Choo, A.D. 420–422 : 3rd year, 422.
> S. D. Heu determined by β Aquarii and another.
> S. D. Wei determined by α Aquarii and θ, ε Pegasi.

144   A.D. 422. *December* 17.

In the 11th moon of the same year, on the day Woo Woo, there was a comet in Ying Shih.

> 422, 11th moon, day Woo Woo, December 17th.
> Yung Shih, same as S. D. Shih, determined by α and others in Pegasus.

145   A.D. 423. *February* 13.

In the reign of Shaou Te, the 1st year of the epoch King Ping, the 1st moon, day Yih Maou, there was a comet in the eastern part of S. D. Peih.

> Emperor Shaou Te and epoch King Ping, A.D. 423 : 1st moon, day Yih Maou, February 13th.
> S. D. Peih determined by γ Pegasi and α Andromedæ.   *She Ke, M. T. L.*

146   A.D. 423. *October* 15.

In the 10th moon of the same year, on the day Ke Wei, there was a comet in S. D. Te.

> 423, 10th moon, day Ke Wei, October 15th.
> S. D. Te determined by α, β, γ, ν Libræ.   *She Ke, M. T. L.*

147   A.D. 442. *November* 1.

In the reign of Wan Te, the 19th year of the epoch Yuen Kea, the 9th moon, day Ping Shin, there was a strange star in Pih Tow. It became a comet, and entered Wan Chang, Kwan and Woo Chay. It swept S. D. Peih. It passed near Teen Tsze. It crossed Teen Yuen. In the winter it disappeared.

> Emperor Wan Te and epoch Yuen Kea, A.D. 424–453 : 19th year, A.D. 442 : 9th moon, day Ping Shin, November 1st.
> S. D. Peih determined by α, γ, δ and others in Taurus.
> Pih Tow, the seven bright stars in Ursa Major.
> Kwan or Shih, Corona Borealis. Wan Chang, θ, φ, ν Ursæ Majoris.
> Woo Chay, α, β, θ, κ Aurigæ and β Tauri.
> Teen Tsze, π, ρ and others in Taurus, near the Hyades.
> Teen Yuen, γ, δ, ε and others in Eridanus.   *She Ke, M. T. L.*

148   A.D. 449. *November* 11.

In the 26th year of the same epoch, 10th moon, day Kwei Maou, a comet was seen in Tae Wei.

> Yuen Kea, 26th year, 449: 10th moon, day Kwei Maou, November 11th.
> Tae Wei, space between stars in Leo and Virgo.   *She Ke, M. T. L.*

149   A.D. 451. *May* 17.

In the 28th year of the same epoch, the 4th moon, day Yih Maou, a comet was seen in S. D. Maou. In the 6th moon, day Jin Tsze, it was seen in the middle of Tae Wei, over against Te Tso.

> Yuen Kea, 28th year, A.D. 451: 4th moon, day Yih Maou, May 17th: 6th moon, day Jin Tsze, July 13th.
> S. D. Maou determined by the Pleiades.
> Tae Wei, space between stars in Leo and Virgo.
> Te Tso, β Leonis and small stars near.   *She Ke, M. T. L.*
> The 'She Ke' has the day Ke Maou, June 10th.

TSE DYNASTY, A.D. 479–501.

150   A.D. 501. *February* 13.

In the reign of Tung Hwan How, in the 3rd year of the epoch Yung Yuen, 1st moon, day Yih Sze, a tailed star was seen in the horizon.

> Emperor Tung Hwan How and epoch Yung Yuen, 499–500: 3rd year, 501: 1st moon, day Yih Sze, February 13th.
> In the Tables, 501 is in the next epoch, Chung Hing.

151   A.D. 501. *April* 14.

In the reign of Ho Te, the 1st year of the epoch Chung Hing, 3rd moon, day Yih Sze, there was a comet in the horizon.

> Ho Te and epoch Chung Hing, A.D. 501: 3rd moon, day Yih Sze, April 14th.
> This and the preceding are possibly the same comet: they are both from 'M. T. L.'

LEANG DYNASTY, A.D. 502–556.

152   A.D. 532. *January* 6. (?)

In the reign of Woo Te, the 5th year of the epoch Chung Ta Tung, 1st moon, day Ke Yew, a tailed star was seen.

> Emperor Woo Te, A.D. 502–549; epoch Chung Ta Tung, 528–534: 5th year, 532; 1st moon, day Ke Yew, January 16th. This date is doubtful.

K

153   A.D. *539. November* 17.

In the 5th year of the epoch Ta Tung, 10th moon, day Sin Chow, a comet appeared in Nan Tow. It was about one cubit in length, pointing to the south-east. It gradually increased to about 10 cubits in length. In the 11th moon, day Yih Maou, it entered S. D. Lew and disappeared.

Epoch Ta Tung, 535–545, 5th year, 539: 10th moon, day Sin Chow, November 17th; 11th moon, day Yih Maou, December 1st.

S. D. Nan Tow, or Tow, determined by ζ, τ, σ, &c. Sagittarii.

S. D. Lew determined by a, β, γ Arietis.   *She Ke, M. T. L.*

CHIN DYNASTY, A.D. 557–588.

154   A.D. *560. October* 4.

In the reign of Wan Te, the 1st year of the epoch Teen Kea, the 9th moon, on the day Kwei Chow, a comet was seen. It was 4 cubits in length. The tail pointed to the south-west.

Emperor Wan Te, A.D. 560–566; epoch Teen Kea, 560–565, 1st year, 560: 9th moon, day Kwei Chow, October 4th.   *She Ke, M. T. L.*

155   A.D. *565. July* 23.

In the 6th year of the same epoch, the 6th moon, day Sin Yew, there was a comet about 10 cubits in length. It was seen in Shang Tae.

Teen Kea, 6th year, A.D. 565: 6th moon, day Sin Yew, July 23rd.

Shang Tae, ι, κ Ursæ Majoris.   *She Ke, M. T. L.*

156   A.D. *568. August* 3.

In the reign of Fei Te, the 2nd year of the epoch Kwang Ta, the 6th moon, day Ting Hae, a comet was seen.

Emperor Fei Te, A.D. 567–568; epoch Kwang Ta, the same; 2nd year, 568: 6th moon, day Ting Hae, August 3rd.   *M. T. L.*

157   A.D. *575. April* 27.

In the reign of Seuen Te, in the 7th year of the epoch Ta Keen, 4th moon, day Ping Seuh, there was a comet near Ta Keo.

Emperor Seuen Te, A.D. 569–582; epoch Ta Keen the same; 7th year, 575: 4th moon, day Ping Seuh, April 27th.

Ta Keo, Arcturus.

158 A.D. 416. *January* 26.

In the 12th year of the same epoch, the 12th moon, day Sin Sze, a comet was seen in the south-west.

Ta Keen, 12th year, 580: 12th moon, day Sin Sze, January 26th.

She Ke, M. T. L.

How Wei, a Minor Dynasty, a.d. 386–534.

159 A.D. 416. *June* 27.

In the reign of Ming Yuen Te, in the 1st year of the epoch Tae Chang, 5th moon, day Kea Shin, two comets were seen.

Ming Yuen Te, A.D. 409–423; epoch Tae Chang, 416–423; 1st year, 416: 5th moon, day Kea Shin, June 27th. *M. T. L.*

Pih Tse, a Minor Dynasty, a.d. 570–577.

160 A.D. 565. (?) *April* 21.

In the reign of Woo Ching Te, the 4th year of the epoch Ho Tsing, 3rd moon, a comet was seen.

Emperor Woo Ching Te, A.D. 561–564; epoch Ho Tsing, 562–564; 4th year, 565 (?): 3rd moon, April. *She Ke, M. T. L.*

The 'She Ke' adds the day, Woo Tsze, April 21.

The Tables give but 3 years to this epoch: the date is consequently doubtful.

161 A.D. 565. *July* 24.

In the reign of How Choo, the 1st year of the epoch Teen Tung, the 6th moon, day Jin Seuh, a comet was seen in Wan Chang. Its length was reckoned at $\frac{1}{10}$th of a cubit. It entered Wan Chang. It passed over Shang Tseang, and afterwards crossed Tsze Wei Kung to its western boundary. It gradually lengthened to about 10 cubits. It pointed to S. D. Shih and Peih. After about 100 days it entered S. D. Heu and Wei, and then disappeared.

Emperor How Choo, A.D. 565–576; epoch Teen Tung, 565–569; 1st year, 565: 6th moon, day Jin Seuh, July 24th.

S. D. Wei determined by α Aquarii and θ, ε Pegasi.

Shih determined by α Pegasi and others near.

Peih determined by γ Pegasi and α Andromedæ.

Heu determined by β Aquarii and others.

Wan Chang, θ, ν, φ, &c. Ursæ Majoris.

Shang Tseang, stars in Coma Berenices.

Tsze Wei Kung, circle of perpetual apparition. *She Ke, M. T. L.*

162   A.D. *568. July.*

In the 4th year of the same epoch, the 6th moon, a comet was seen in S. D. Tsing.
    Teen Tung, 4th year, A.D. 568 : 6th moon, July.
    S. D. Tsing, γ, ε, λ, μ, &c. Geminorum.   *She Ke, M. T. L.*

163   A.D. *568. August.*

In the 7th moon of the same year a comet was seen in S. D. Fang and Sin. It
was white like meal, or the refuse of silk, and was as large as a tow measure. It went
to the east. In the 8th moon it entered Teen She. It gradually increased in length to
40 cubits. In shape it resembled a melon. It passed through S. D. Heu and Wei. It
entered S. D. Shih. It passed over the Le Kung. In the 9th moon it entered S. D.
Kwei. It passed on to S. D. Lew, and then disappeared.

    Teen Tung, 4th year : 7th moon, 568, August; 8th moon, September; 9th
moon, October.
        S. D. Fang determined by β, δ, π, ρ in Scorpio.
            Sin determined by α, σ, τ in Scorpio.
            Heu determined by β Aquarii and another.
            Wei determined by α Aquarii and θ, ε Pegasi.
            Shih determined by α, β Pegasi and others near.
            Kwei determined by β, δ, ε Andromedæ and stars in Pisces.
            Lew determined by α, β, γ Arietis, &c.
    Teen She, space bounded by Serpens.
    Le Kung, three groups of two stars each in Pegasus : they are λ μ, η o, ν τ,
and form part of S. D. Shih.   *She Ke, M. T. L.*

How Chow, a Minor Dynasty, A.D. 557–581.

164   A.D. *561. September 26.*

In the reign of Woo Te, the 1st year of the epoch Paou Ting, the 9th moon, day
Yih Sze, an extraordinary star was seen in S. D. Yih.

    Emperor Woo Te, 561–578; epoch Paou Ting, 561–565; 1st year, 561 : 9th
moon, day Yih Sze, September 26th.
    S. D. Yih determined by α and others in Crater.   *She Ke.*

165   A.D. *565. July 22.*

In the 5th year of the same epoch, the 6th moon, day Kang Shin, a comet appeared
in San Tae. It entered Wan Chang and Shang Tseang. It afterwards crossed the
western boundary of Tsze Kung. It entered S. D. Wei, and gradually increased to
about 10 cubits in length. It pointed towards S. D. Shih and Peih. After about 100
days it gradually diminished to about 2½ cubits in length. It arrived at S. D. Heu and
Wei, and then disappeared.

Paou Ting, 5th year, A.D. 565 : 6th moon, day Kang Shin, July 22.
S. D. Wei determined by a Aquarii and θ, ε Pegasi.
Shih determined by a, β Pegasi and stars near.
Peih determined by a Tauri and others near.
Heu determined by β Pegasi and another.
San Tae, feet of Ursa Major.
Wan Chang, θ, ν, ϕ, &c. Ursæ Majoris.
Shang Tseang, ν, &c. in Coma Berenices.
Tsze Kung, circle of perpetual apparition. *M. T. L.*

This appears to be the same as No. 161 by a different observer, and on another day.

166   A.D. 568. *July* 21.

In the 3rd year of the epoch Teen Ho, 6th moon, day Kea Seuh, a comet was seen in the eastern part of S. D. Tsing. It was 10 cubits in length. Its colour was white in the upper part and reddish below. It ended in a point. It gradually went to the east. In the 7th moon, day Kwei Maou, it passed to the north of S. D. Kwei. It was then $\frac{8}{10}$ths of a cubit in length. It afterwards disappeared.

Epoch Teen Ho, A.D. 566–571, 3rd year, 568 : 6th moon, day Kea Seuh, July 21st; 7th moon, day Kwei Maou, August 19th.
S. D. Tsing determined by γ, δ, λ, μ, &c. Geminorum.
S. D. Kwei determined by γ, δ, η, θ Cancri. *She Ke, M. T. L.*

167   A.D. 574. *April* 4.

In the 3rd year of the epoch Keen Tih, the 2nd moon, day Woo Woo, a strange star, resembling a large peach, of a bluish-white colour, appeared in Woo Chay, to the south-east. It was 3 cubits in length. It went slowly to the east, and whilst there increased to 2 cubits in length. In the 4th moon, day Jin Shin, it entered Wan Chang. On the day Ting Wei it entered Kwei in Pih Tow, to about the middle. It afterwards left Kwei, and gradually became smaller. It was visible altogether for 93 days.

She Ke.

Epoch Keen Tih, A.D. 572–577, 3rd year, 574 : 2nd moon, day Woo Woo, April 4th; 4th moon, day Jin Shin, May 8th; Ting Wei, May 23rd.
Kwei in Pih Tow, the middle of the square in Ursa Major.
Wan Chang, θ, ν, ϕ Ursæ Majoris.

168   A.D. 574. *May* 31.

In the same year, the 4th moon, day Yih Maou, there was a comet just without the boundary of Tsze Kung. It was large, like a man's fist : colour, reddish white. It pointed to Woo Te Tso. It went slowly to the south-east. Its length was 15 cubits. In the 5th moon, day Kea Tsze, it went to the north of Shang Tae and disappeared.

L

Keen Tih, 3rd year, 574: 4th moon, day Yih Maou, May 31st; 5th moon, day Kea Tsze, June 9th.

Tsze Kung, circle of perpetual apparition.

Woo Te Tso, β Leonis and small stars near.

Shang Tae, ι, κ Ursæ Majoris.   *M. T. L.*

In the 'She Ke' this is placed in the 10th moon.

Suy Dynasty, A.D. 589-617.

169   A.D. *588. November* 22.

In the reign of the Emperor Wan Te, the 8th year of the epoch Kae Hwang, the 10th moon, day Kea Tsze, there was a comet in Keen New.

Wan Te, one of the minor Princes, assumed the Imperial title, and thus became the founder of the Suy dynasty, A.D. 589, which was the 9th year of his epoch Kae Hwang, 581–600: consequently the 8th year was 588. His reign closed in 604.

10th moon, day Kea Tsze, November 22nd.

Keen Neu, same as S. D. Neu, determined by α, β, &c. Capricorni.

She Ke, M. T. L.

170   A.D. *594. November* 10.

In the 14th year of the same epoch, the 11th moon, day Kwei Wei, there was a comet in S. D. Heu and Wei. It extended to S. D. Kwei and Lew.

Kae Hwang, 14th year, 594: 11th moon, day Kwei Wei, November 10th.

S. D. Heu determined by β Aquarii and another.

Wei determined by α Aquarii and θ, ε Pegasi.

Kwei determined by β, δ, ε, &c. Andromedæ and stars in Pisces.

Lew determined by α, β, γ Arietis.   *She Ke, M. T. L.*

171   A.D. *607. March* 13.

In the reign of Yang Te, the 3rd year of the epoch Ta Nae, 2nd moon, day Ke Chow, a comet was seen in the eastern part of S. D. Tsing and Wan Chang. It passed through Ta Ling, Woo Chay, and Pih Ho. It entered Tae Wei and swept Te Tso. It passed on, and after about 100 days it disappeared.

Emperor Yang Te and epoch Ta Nae, A.D. 605–616, 3rd year, 607: 2nd moon, day Ke Chow, March 13th.

S. D. Tsing determined by γ, ε, λ, μ Geminorum.

Wan Chang, θ, ν, φ, &c. Ursæ Majoris.   Ta Ling, τ and others in Perseus.

Woo Chay, α, β, θ, χ Aurigæ and β Tauri.   Pih Ho, α, β, &c. Geminorum.

Tae Wei, space between stars in Leo and Virgo.

Te Tso, β Leonis and stars near.   *M. T. L.*

172   A.D. 607. *April 4.*

In the 3rd moon of the same year, day Sin Hae, a tailed star was seen in the horizon to the west. It passed through S. D. Kwei, Lew, Keo, and Kang, and then was no longer seen. In the 9th moon, on the day Sin Wei, it returned, and was seen in the south. It was of a reddish colour, and was in the horizon in S. D. Keo and Kang, near their boundaries. It swept Tae Wei near Te Tso. It entered most of the S. D., but did not extend to Tsan and Tsing. In the beginning of the next year it disappeared.

Ta Nae, 3rd year, A.D. 607: 3rd moon, day Sin Hae, April 4th; 9th moon, day Sin Wei, October 21st.

S. D. Kwei determined by α, δ, ε, &c. Andromedæ and stars in Pisces.
Lew determined by α, β, γ Arietis.
Keo determined by α and ζ Virginis.
Kang determined by ι, κ, λ, μ Virginis.
Tsan determined by α, β, &c. Orionis.
Tsing determined by γ, ε, μ, &c. Geminorum.
Tae Wei, space between stars in Leo and Virgo.
Te Tso, β Leonis and other stars near.   *She Ke, M. T. L.*

This may relate to two comets, as the account is not very clear.

173   A.D. 615. *July.*

In the 11th year of the same epoch, the 6th moon, there was a comet in Wan Chang, to the south-east. Its length was from 5 to 6 tenths of a cubit. Its colour was dusky, and its extremity pointed. In the evening it had a waving motion. It went to the north-west. For several days it was in Wan Chang. It went within 4 or 5 tenths of a cubit of Kung, but did not enter that space, and disappeared.

Ta Nae, 11th year, 615: 6th moon, July.
Wan Chang, θ, φ, ν Ursæ Majoris.
Kung, or Tsze Kung, circle of perpetual apparition.   *M. T. L.*

174   A.D. 616. *July.*

In the 13th year of the same epoch, the 6th moon, there was a comet in Tae Wei, near Woo Te Tso. Its colour was a yellowish red. It was from 3 to 4 cubits in length. After several days it disappeared.

Ta Nae, 13th year, A.D. 616: 6th moon, July.
Tae Wei, space between stars in Leo and Virgo.
Woo Te Tso, β Leonis and small stars near.   *M. T. L.*

175   A.D. 616. *October.*

In the 9th moon of the same year a comet was seen in Yung Shih.

616: 9th moon, October.
Yung Shih, same as S. D. Shih, determined by α Pegasi and other stars near.   *M. T. L.*

TANG DYNASTY, A.D. 618–906.

176   A.D. 626. *March 26.*

In the reign of Kaou Tsoo, the 9th year of the epoch Woo Tih, the 2nd moon, day Jin Woo, there was a comet in the S. D. Wei and Maou. On the day Ting Hae the comet was in Keuen She.

> Emperor Kaou Tsoo, A.D. 618–626; epoch Woo Tih the same: 9th year, 626: 2nd moon, day Jin Woo, March 26th; Ting Hae, March 31st.
> S. D. Wei determined by the three stars in Musca.
> S. D. Maou, determined by the Pleiades.
> Keuen She, ν Persei.   *She Ke, M. T. L.*

177   A.D. 634. *September 22.*

In the reign of Tae Tsung, the 8th year of the epoch Ching Kwan, the 8th moon, day Kea Tsze, there was a comet in S. D. Heu and Wei. It passed through Heuen Heaou. On the day Yih Hae it was no longer visible.

> Emperor Tae Tsung and epoch Ching Kwan, A.D. 627–649, 8th year, 634: 8th moon, day Kea Tsze, September 22nd; Yih Hae, October 3rd.
> S. D. Heu determined by β Aquarii and another.
> S. D. Wei determined by α Aquarii and θ, ε Pegasi.
> Heuen Heaou, one of the 12 kung, answering to our sign Aquarius, and comprising S. D. Neu, Heu, and Wei.   *M. T. L.*

178   A.D. 639.

In the 13th year of the same epoch, the 3rd moon, day Yih Chow, there was a comet in S. D. Peih and Maou.

> 13th year of Ching Kwan, 639: 3rd moon, day Yih Chow, April 30th.
> S. D. Peih determined by α, γ, δ, &c. Tauri.
> S. D. Maou determined by the Pleiades.   *She Ke, M. T. L.*
> The 'She Ke' makes the year 638.

179   A.D. 641. *August 1.*

In the 15th year of the same epoch, 6th moon, day Ke Yew, there was a comet in Tae Wei. It passed over Lang Wei. In the 7th moon, day Kea Seuh, it was no longer visible.

> Ching Kwan, 15th year, A.D. 641: 6th moon, day Ke Yew, August 1st; 7th moon, day Kea Seuh, August 26th.
> Tae Wei, space between stars in Leo and Virgo.
> Lang Wei, Coma Berenices.   *She Ke, M. T. L.*

180 A.D. 663. *September* 29.

In the reign of Kaou Tsung, the 3rd year of the epoch Lung So, 8th moon, day Kwei Maou, there was a comet in Tso She Te. It was about 2 cubits in length. On the day Yih Sze it was no longer visible.

Emperor Kaou Tsung, A.D. 650–683; epoch Lung So, 661–663, 3rd year, 663 : 8th moon, day Kwei Maou, September 29th ; day Yih Sze, October 1st.

Tso She Te, ξ, o, π Boötis. *She Ke, M. T. L.*

181 A.D. 667. *May* 24.

In the 2nd year of the epoch Keen Fung, 4th moon, day Ping Shin, there was a comet in the north-east. Its place was in Woo Chay, between S. D. Peih and Maou. On the day Yih Hae it was no longer visible.

Epoch Keen Fung, 666–667 : 2nd year, 667 : 4th moon, day Ping Shin, May 24th ; day Yih Hae, June 12th.

S. D. Peih determined by a, γ, δ, &c. Tauri.

S. D. Maou determined by the Pleiades.

Woo Chay, a, β, θ, κ Aurigæ and β Tauri. *She Ke, M. T. L.*

182 A.D. 676. *January* 3.

In the 2nd year of the epoch Shang Yuen, the 12th moon, day Jin Woo, there was a comet to the south of S. D. Keo and Kang. It was 5 cubits in length.

Epoch Shang Yuen, 674–675 : 2nd year, 675 : 12th moon, day Jin Woo, 676, January 3rd.

S. D. Keo determined by a and ζ Virginis.

S. D. Kang determined by ι, κ, λ, θ Virginis. *She Ke, M. T. L.*

183 A.D. 676. *July* 7.

In the 3rd year of the same epoch, the 7th moon, day Ting Hae, there was a comet in the eastern part of S. D. Tsing, pointing towards Pih Ho. It was about 3 cubits in length. Its luminous envelope increased greatly until it became 30 cubits in length. It pointed towards Chung Tae and Wan Chang. In the 9th moon, day Yih Yew, it disappeared.

Epoch Shang Yuen, 3rd year, 676 : 7th moon, day Ting Hae, July 7th ; 9th moon, day Yih Yew, September 3rd.

S. D. Tsing determined by δ, ε, λ, μ, &c. Geminorum.

Pih Ho, a, β, ρ, σ Geminorum.

Chung Tae, λ, μ Ursæ Majoris.

Wan Chang, θ, ν, φ Ursæ Majoris. *She Ke, M. T. L.*

The 'She Ke' has 'swept Chung Tae and Wan Chang,' which appears to be the preferable reading.

M

184   A.D. 681. *October* 17.

In the 1st year of the epoch Kae Tih, the 9th moon, day Ping Shin, there was a comet in the middle of Teen She. It was 50 cubits in length. It gradually lessened and went to the east. It passed on to Ho Koo. On the day Kwei Chow it was no longer visible.

> Epoch Kae Tih, 1st year, A.D. 681 : 9th moon, day Ping Shin, October 17th : day Kwei Chow, November 3rd.
> Teen She, space bounded by Serpens.
> Ho Koo, *a*, *β*, *γ* Aquilæ.   *She Ke, M. T. L.*

185   A.D. 683. *April* 20.

In the 2nd year of the epoch Yung Shun, the 3rd moon, day Ping Woo, there was a comet to the north of Woo Chay. In the 4th moon, day Sin Wei, it disappeared.

> Epoch Yung Shun, A.D. 682–683 : 2nd year, 683 : 3rd moon, day Ping Woo, April 20th ; day Sin Wei, May 15th.
> Woo Chay, *a*, *β*, &c. Aurigæ and *β* Tauri.   *She Ke, M. T. L.*

186   A.D. 684. *July* 8.

In the epoch Wan Ming, 1st year, 7th moon, day Sin Wei, there was a comet in the west. It was about 10 cubits in length. In the 8th moon, day Kea Shin, it disappeared.

> The epoch Wan Ming does not occur as one of those of this dynasty. In the 'Tung Keen Kang Muh,' vol. lv., it is mentioned as that of one of the princes who assumed sovereignty about this time, and the 1st year coincides with the 1st year of the Emperor Chung Tsung : hence it is A.D. 684.
> 7th moon, day Sin Wei, July 8th ; 8th moon, day Kea Shin, August 10th.
>   *She Ke, M. T. L.*
> Biot makes this September 6th and October 9th ; by computation it comes out as I have rendered it.

187   A.D. 684. *September* 12.

In the reign of Chung Tsung, the 1st year of the epoch Kwang Tsih, the 9th moon, day Ting Chow, there was a star resembling a half moon in the west.

> Emperor Chung Tsung, A.D. 684–709. The epoch Kwang Tsih is not in the regular list. In ' M. T. L.' it is the 1st epoch of Chung Tsung, and this is, there-fore, his 1st year, 684 : 9th moon, day Ting Chow, September 12th. Biot makes it October 11th.

> This was most likely a meteor.

188   A.D. 707. *November* 16.

In the 1st year of the epoch King Lung, the 10th moon, day Jin Woo, there was a comet in the west. In the 11th moon, day Kea Yin, it disappeared.

> King Lung, A.D. 707–709: 1st year, A.D. 707: 10th moon, day Jin Woo, November 16th; 11th moon, day Kea Yin, December 17th.   *She Ke, M. T. L.*

189   A.D. 708. *March* 30.

In the 2nd year of the same epoch, 2nd moon, day Ting Yew, there was a comet between S. D. Wei and Maou.

> King Lung, 2nd year, 708 : 2nd moon, day Ting Yew, March 30th.
> S. D. Wei determined by three stars in Musca.
> S. D. Maou determined by the Pleiades.   *She Ke, M. T. L.*

190   A.D. 708. *September* 21.

In the 8th moon of the same year, day Jin Shin, there was a comet in Tsze Kung.

> 8th moon, day Jin Shin, September 21st.
> Tsze Kung, circle of perpetual apparition.

191   *Between* A.D. 710 *and* A.D. 713.

In the 1st year of the epoch Yen Ho, the 6th moon, there was a comet. From Heen Yuen it entered Tae Wei. It passed on to Ta Keo and disappeared.

> The epoch Yen Ho is not one of the regular epochs of this dynasty. It appears to have been somewhere between 710 and 713.
> Heen Yuen, Regulus and other stars in Leo and Leo Minor.
> Ta Wei, space between stars in Leo and Virgo.
> Ta Keo, Arcturus.

192   A.D. 730. *June* 30.

In the reign of Yuen Tsung, the 18th year of the epoch Kae Yuen, the 6th moon, day Kea Tsze, there was a comet in Woo Chay. On the day Kwei Yew the comet was in S. D. Peih and Maou.

> Emperor Yuen Tsung, called also Heuen Tsung, A.D. 713–755; epoch Kae Yuen, 713–741: 18th year, 730: 6th moon, day Kea Tsze, June 30th; day Kwei Yew, July 9th.
> S. D. Peih determined by a, γ, δ, ε Tauri.
> S. D. Maou determined by the Pleiades.
> Woo Chay, a, β, γ Aurigæ and β Tauri.

> The latter portion of this, from Kwei Yew, is separate in the original, both in the 'She Ke' and 'M. T. L.' It is, however, evident that both relate to the same comet.

193   A.D. 739.

In the 26th year of the same epoch, the 3rd moon, day Ping Tsze, there was a comet in Tsze Kung. It was bright. It passed through Pih Tow Kwei. After 10 days, being obscured by clouds, it was no more seen.

> Kae Yuen, 26th year, 739.
> Pih Tow Kwei, the square in Ursa Major.   *She Ke, M. T. L.*

194   A.D. 760. *May 16.*

In the reign of Suh Tsung, the 3rd year of the epoch Keen Yuen, the 4th moon, day Ting Sze, there was a comet in the east. Its place was between S. D. Lew and Wei. Its colour was white. It was 4 cubits in length. It went rapidly to the east. It passed through S. D. Maou, Peih, Tsuy He, Tsan, and Tung Tsin, to Kwei, Lew, and Heen Yuen. It passed to the west of Yew Chih Fa. It was seen altogether for about 50 days.

> Emperor Suh Tsung, A.D. 756–762; epoch Keen Yuen, 758–759 : 3rd year, day Ting Sze, 760, May 16th.
> The Tables give but two years to the epoch Keen Yuen.
> S. D. Lew determined by a, β, γ Arietis.
>> Wei determined by the three stars in Musca.
>> Maou determined by the Pleiades.
>> Peih determined by a, γ, δ, ϵ, &c. Tauri.
>> Tsuy He, or Tsuy, determined by λ and stars in head of Orion.
>> Tsan determined by a, β, γ, δ Orionis.
>> Kwei determined by γ, δ, η, θ Cancri.
>> Lew determined by δ, &c. Hydræ.
>> Tsing determined by γ, ϵ, λ, μ Geminorum.
> Heen Yuen, a Leonis and others in Leo and Leo Minor.
> Yew Chih Fa, β Virginis.   *She Ke, M. T. L.*

195   A.D. 760. *May 15.*

In the intercalary moon of the same year, on the day Sin Yew, the 1st day of the moon, a comet was seen in the west. It was 10 cubits in length. When the 5th moon commenced it had disappeared.

> The intercalary moon appears to have been that which preceded the 5th moon. The day Sin Yew will, therefore, be May 15, and the 5th moon June or July.
>   *She Ke, M. T. L.*

196   A.D. 767. *January 12.*

In the reign of Tae Tsung, the 1st year of the epoch Ta Leih, the 12th moon, day Ke Hae, there was a comet in Kwa Chaou. It was about a cubit in length. After 20 days it disappeared. It passed over Hwan Chay.

Emperor Tae Tsung, A.D. 763–769; epoch Ta Leih, 766–779: 1st year, 766:
12th moon, day Ke Hae, January 12, 767.

Kwa Chaou, a, β, γ, &c. Delphini.

Hwan Chay, ε, ι, &c. Ophiuchi.   *She Ke, M. T. L.*

197   A.D. 770. *June* 15.

In the 5th year of the same epoch, the 4th moon, day Ke Wei, there was a comet
in Woo Chay. Its luminous envelope appeared much disordered. It was about 30
cubits in length. In the 5th moon, day Ke Maou, the comet was seen in the north.
Its colour was white. On the day Kwei Wei it went to the east, and approached the
middle star of Pa Kuh. In the 6th moon, day Kwei Maou, it came near San Kung.
On the day Ke Wei it disappeared.

Epoch Ta Leih, 5th year, 770: 5th moon, day Ke Maou, June 15th: days,
Kwei Wei, June 19th; Kwei Maou, July 9th; Ke Wei, July 25th.

San Kung, three stars near head of Asterion.

Pa Kuh, δ, ξ Aurigæ.   *She Ke, M. T. L.*

197[*]   A.D. 773. *January* 17.

In the 7th year of the same epoch, the 12th moon, day Ping Yin, there was a tailed
star in the lower part of S. D. Tsan. The tail of this comet extended across the heavens
from the star Tang in S. D. Tsan.

7th year of epoch Ta Leih, 772: 12th moon, day Ping Yin, 773, January 17.

S. D. Tsan determined by a and other stars in Orion.

Tang, a star in Orion not identified.   *She Ke, M. T. L.*

198   A.D. 815. *April.*

In the reign of Heen Tsung, the 10th year of the epoch Yuen Ho, there was a
tailed star in Tae Wei. The tail extended to Heen Yuen.

Emperor Heen Tsung, A.D. 806–820; epoch Yuen Ho, the same; 10th year,
A.D. 815: 3rd moon, April.

Tae Wei, space between stars in Leo and Virgo.

Heen Yuen, a, γ, ε, λ and others in Leo and Leo Minor.   *She Ke, M. T. L.*

199   A.D. 817. *February* 17.

In the 12th year of the same epoch, the 1st moon, day Woo Tsze, there was a
comet in S. D. Peih.

Yuen Ho, 12th year, 817: 1st moon, day Woo Tsze, February 17th.

S. D. Peih determined by a and others in Taurus.   *She Ke, M. T. L.*

N

200   A.D. 821. *February 27.*

In the reign of Muh Tsung, the 1st year of the epoch Chang King, 1st moon, day Ke Wei, there was a comet in S. D. Yih.

> Emperor Muh Tsung and epoch Chang King, 821–824, 1st year: 1st moon, day Ke Wei, 821, February 27th.
> S. D. Yih determined by *a* and others in Crater.   *She Ke, M. T. L.*

201   A.D. 821. *March 7.*

In the 2nd moon of the same year, day Ting Maou, there was a comet in Tae Wei, to the west of the star Shang Tseang. In the 6th moon the comet was in S. D. Maou. Its length was 10 cubits. It was visible altogether for 10 days, after which it disappeared.

> 2nd moon, day Ting Maou, 821, March 7 : 6th moon, July.
> Tae Wei, space between stars in Leo and Virgo.
> Shang Tseang, σ Leonis.
> S. D. Maou determined by the Pleiades.   *She Ke, M. T. L.*

202   A.D. 828. *July 5.*

In the reign of Wan Tsung, the 2nd year of epoch Tae Ho, the 7th moon, day Kea Shin, there was a comet in Yew She Te, to the south. Its length was 2 cubits.

> Emperor Wan Tsung, A.D. 827–840; epoch Tae Ho, 827–835: 2nd year, 828 : 7th moon, day Kea Shin, July 5.
> Yew She Te, η, υ, τ in Boötes.   *She Ke, M. T. L.*

203   A.D. 829. *December.*

In the 3rd year of the same epoch, the 10th moon, a strange star was seen in Shwuy Wei.

> Tae Ho, 3rd year, 829 : 10th moon, November.
> Shwuy Wei, ζ, θ, o Canis Minoris.   *She Ke.*

204   A.D. 834. *October 9.*

In the 8th year of the same epoch, the 9th moon, day Sin Hae, there was a comet in Tae Wei. It was about 10 cubits in length. Its course was to the north-west. It passed over Lang Wei. On the day Kang Shin it was no longer visible.

> Tae Ho, 8th year, 834 : 9th moon, day Sin Hae, October 9th ; Kang Shin, October 18th.
> Tae Wei, space between stars in Leo and Virgo.
> Lang Wei, Coma Berenices.   *She Ke, M. T. L.*

205   A.D. 837. *March* 22.

In the 2nd year of the epoch Kae Ching, the 2nd moon, day Ping Woo, there was comet in S. D. Wei. It was about 7 cubits in length. It pointed towards Nan Tow. On the day Woo Shin it was to the south-west of S. D. Wei. It was bright, and moved rapidly. On the day Kwei Chow its place was in S. D. Heu. On the day Sin Yew its length was about 10 cubits. It went to the west, gradually pointing to the south. On the day Jin Seuh its place was in Woo Neu: its length was about 20 cubits, and was 3 cubits in breadth. On the day Kwei Hae the tail was still broad. In the 3rd moon, day Kea Tsze, its place was in Nan Tow. On the day Yih Chow its length was 50 cubits, the end (of the tail) being divided into two branches, the one pointing to S. D. Te, the other covering S. D. Fang. On the day Ping Yin its length was 6 cubits, and was no longer branched. It pointed to the north. Its place was in the 7th degree of S. D. Kang. On the day Ting Maou it went to the north-west, pointing to the east. On the day Ke Sze its length was about 80 cubits: its place was then in S. D. Chang. On the day Kwei Wei it was but 3 cubits in length: its place was to the right of Heen Yuen. After this it was no longer visible.

Epoch Kae Ching, A.D. 836–840: 2nd year, 837: 2nd moon, day Ping Woo, March 22: days, Woo Shin, March 24; Kwei Chow, March 29; Sin Yew, April 6; Jin Seuh, April 7; Kwei Hae, April 8: 3rd moon, day Kea Tsze, April 9; Yih Chow, April 10; Ping Yin, April 11; Ting Maou, April 12; Ke Sze, April 14; Kwei Wei, April 28.

    S. D. Wei determined by a Aquarii and θ, ϵ Pegasi.
        Heu determined by β Aquarii and another.
        Te determined by a, β, γ, ι Libræ.
        Fang determined by β, δ, π in Scorpio.
        Kang determined by ι, κ, λ, θ Virginis.
        Chang determined by κ, λ, μ, &c. Hydræ.
        Tow, or Nan Tow, determined by δ, μ, &c. Sagittarii.
        Woo Neu, or Neu, determined by ϵ, μ, ν, &c. Aquarii.
    Heen Yuen, a and others in Leo and Leo Minor.

'M. T. L.' adds, in a kind of note, a sentence implying that, generally speaking, it may be looked upon as a constant rule, that when a comet appears in the morning its direction is to the west, and to the east when it appears in the evening.

206   A.D. 837. *April* 29.

In the same moon, day Kea Shin, a strange star was seen in the lower part of S. D. Tsing, to the east.

    Day Kea Shin, 837, April 29th.
    S. D. Tsing determined by γ, ϵ, λ, μ, &c. Geminorum.   *She Ke.*

207   A.D. 837. *May 3.*

On the day Woo Tsze a strange star was seen within Twan Mun, near the star Ping.

    Day Woo Tsze, May 3rd.
    Twan Mun, between β and η Virginis.
    Star Ping, ν and others in Virgo.   *She Ke.*

208   A.D. 837. *May 21.*

In the 4th moon of the same year, day Ping Woo, the strange star seen in the lower part of S. D. Tsing, to the east, disappeared.

    837: 4th moon, day Ping Woo, May 21st.
    S. D. Tsing, as above. This relates to No. 206.   *She Ke.*

209   A.D. 837. *June 17.*

In the 5th moon, day Kwei Yew, the strange star seen in Twan Mun disappeared.

    5th moon, day Kwei Yew, June 17th.
    Twan Mun. See No. 207, to which this relates.   *She Ke.*

210   A.D. 837. *June 26.*

In the same moon, day Jin Woo, a strange star, like a comet, was seen in Nan Tow, near Teen Yo.

    Day Jin Woo, June 26th.
    Nan Tow, same as S. D. Tow, determined by δ, μ, &c. Sagittarii.
    Teen Yo, not identified.   *She Ke.*

211   A.D. 837. *September 9.*

In the 8th moon of the same year, day Ting Yew, there was a comet in the S. D. Heu and Wei.

    837: 8th moon, day Ting Yew, September 9th.
    S. D. Heu determined by β Aquarii and another.
    S. D. Wei determined by α Aquarii and θ, ϵ Pegasi.   *She Ke, M. T. L.*

212   A.D. 838. *November 11.*

In the 3rd year of the same epoch, the 10th moon, day Yih Sze, there was a comet in S. D. Chin and Kwei. It was about 20 cubits in length. The tail gradually pointed to the west.

    Kae Ching, 3rd year, 838: 10th moon, day Yih Sze, November 11th.
    S. D. Chin determined by β, &c. Corvi.
    S. D. Kwei determined by γ, δ, &c. Cancri.   *She Ke, M. T. L.*

213   A.D. 838. *November* 21.

In the 11th moon, day Yih Maou, there was a comet in the east. Its place was in
S. D. Wei and Ke, from east to west. It extended across the heavens. In the 12th
moon, day Jin Shin, it was no longer seen.

   11th moon, day Yih Maou, November 21 ; Jin Shin, December 8.
   S. D. Wei determined by ε, μ, ν, &c. in Scorpio.
   S. D. Ke determined by γ, δ, ε, &c. Sagittarii.   *She Ke, M. T. L.*
   This may possibly be a continuation of the preceding account.

214   A.D. 839. *February* 7.

In the 4th year, 1st moon, day Kwei Yew, there was a comet in Yu Lin.
   Kae Ching, 4th year, 839 : 1st moon, day Kwei Yew, February 7th.
   Yu Lin, δ, τ, χ and others in Aquarius.   *She Ke, M. T. L.*

215   A.D. 839. *March* 12.

In the intercalary moon of the same year, day Ping Woo, there was a comet in
Keuen Che, to the north-west. In the 2nd moon, day Ke Maou, it disappeared.

   The intercalary moon appears to have been that immediately preceding the
2nd moon. Hence the day Ping Woo will be March 12, and the 2nd moon, Ke
Maou, April 14.
   Keuen Che, ε, ν and others in Perseus.   *She Ke, M. T. L.*

216   A.D. 840. *March* 20.

In the 5th year of the same epoch, 2nd moon, day Kang Shin, there was a comet
in Ying Shih, to the east, between that and S. D. Peih. On the 20th day it disappeared.
   Epoch Kae Ching, 5th year, 840 : 2nd moon, day Kang Shin, March 20.
   Ying Shih, same as S. D. Shih, determined by α Pegasi and others.
   S. D. Peih, determined by γ Pegasi, &c.   *She Ke, M. T. L.*

217   A.D. 840. *December* 3.

In the 11th moon of the same year, day Woo Shin, there was a comet in the east.
   840 : 11th moon, day Woo Shin, December 3rd.   *She Ke, M. T. L.*

218   A.D. 841. *July.*

In the reign of Woo Tsung, the 1st year of the epoch Hwuy Chang, the 7th moon,
there was a comet in Yu Lin, between Ying Shih and the east of the S. D. Peih.
   Emperor Woo Tsung and epoch Hwuy Chang, A.D. 841–846 : 1st year, 841 :
7th moon, July.

S. D. Shih determined by *a* Pegasi and others.
Ying Shih, same as *a* Pegasi.
S. D. Peih determined by γ Pegasi and *a* Andromedæ.
Yu Lin, δ, τ and others in Aquarius.   *She Ke, M. T. L.*

219   A.D. 841. *December 22.*

In the 11th moon of the same year, day Jin Yin, there was a comet near Pih Lo Sze Mun. Its place was in Ying Shih. It entered Tsze Kung. In the 12th moon, day Sin Maou, it was no longer visible.

    841 : 11th moon, day Jin Yin, December 22 ; 12th moon, day Sin Maou, February 9, 842.
    S. D. Shih determined by *a* Pegasi and others.
    Pih Lo Sze Mun, Fomalhaut.
    Tsze Kung, circle of perpetual apparition.   *She Ke, M. T. L.*
    This date is unsatisfactory, the day Sin Maou not falling in the 12th moon.

220   A.D. 851. *April.*

In the reign of Seuen Tsung, the 6th year of the epoch Ta Chung, the 3rd moon, there was a comet in S. D. Tsuy and Tsan, near the star Tang.

    Emperor Seuen Tsung and epoch Ta Chung, A.D. 846–859 : 6th year, 851 ; 3rd moon, April.
    S. D. Tsuy determined by λ and small stars in head of Orion.
    S. D. Tsan determined by *a*, β, γ, δ Orionis.
    Tang, unascertained star in Orion.   *She Ke, M. T. L.*

221   A.D. 856. *September 27.*

In the 11th year of the same epoch, the 9th moon, day Yih Wei, there was a comet in S. D. Fang. It was 3 cubits in length.

    Ta Chung, 11th year, 856 : 9th moon, day Yih Wei, September 27.
    S. D. Fang determined by β, δ, π, &c. in Scorpio.   *She Ke, M. T. L.*

222   A.D. 864. *June 21.*

In the reign of the Emperor E Te Tsung, the 5th year of the epoch Han Tung, the 5th moon, day Ke Hae, in the evening, a comet was seen in the north-east, through an opening in the clouds, for not more than 15 minutes. Its colour was yellowish white. It was 3 cubits in length, and was in S. D. Lew.

    Emperor E Te Tsung and epoch Han Tung, A.D. 860–873 : 5th year, 864 ; 5th moon, day Ke Hae, June 21.
    S. D. Lew determined by *a*, β, γ Arietis.   *She Ke, M. T. L.*

223   A.D. 868.   *February.*

In the 9th year of the same epoch, the 1st moon, there was a comet in S. D. Lew and Wei.

   Han Tung, 9th year, 868 : 1st moon, February.
   S. D. Lew determined by *a, β, γ* Arietis.
   S. D. Wei determined by the three stars in Musca.   *She Ke, M. T. L.*

224   A.D. 869.   *September.*

In the 10th year of the epoch Han Tung, 8th moon, there was a comet to the north-east of Ta Ling.

   Han Tung, 10th year, 869 : 8th moon, September.
   Ta Ling, *τ* and others in Perseus.   *She Ke, M. T. L.*

225   A.D. 877.   *June.*

In the reign of He Tsung, the 4th year of the epoch Keen Foo, the 5th moon, a comet was seen.

   Emperor He Tsung, A.D. 874–888 ; epoch Keen Foo, 874–879 : 4th year, 877 :
   5th moon, June.   *She Ke, M. T. L.*

226   A.D. 885.

In the 1st year of the epoch Kwang Ke a comet was seen in Tseih Shwuy, between that and Tseih Sin.

   Epoch Kwang Ke, 885–887 : 1st year, 885.
   Tseih Shwuy, *λ, μ* Persei.
   Tseih Sin, *χ* Geminorum and *μ* Cancri.   *She Ke, M. T. L.*

227   A.D. 886.   *June 13.*

In the 2nd year of the same epoch, 5th moon, day Ping Seuh, there was a comet in S. D. Wei and Ke. It passed through Pih Tow and She Te.

   Kwang Ke, 2nd year, 886 : 5th moon, day Ping Seuh, June 13.
   S. D. Wei determined by *ε, μ, ν,* &c. in Scorpio.
   S. D. Ke determined by *γ, δ, ε* Sagittarii.
   Pih Tow, *a, β, γ,* &c. Ursæ Majoris.
   She Te, stars in feet of Boötis.

228   A.D. 891.   *May 12.*

In the reign of Chaou Tsung, the 2nd year of the epoch Ta Shun, the 4th moon, day Kang Shin, there was a comet in San Tae. It went to the east. It entered Tae Wei. It swept Ta Keo and Teen She. It was about 100 cubits in length. In the 5th moon, day Kea Seuh, it was no longer visible.

Emperor Chaou Tsung, A.D. 889–904; epoch Ta Shun, 890–891 : 2nd year, 891 : 4th moon, day Kang Shin, May 12 : 5th moon, day Kea Seuh, July 5.

San Tae, feet of Ursa Major.
Tae Wei, space within stars in Leo and Virgo.
Teen She, space bounded by Serpens.
Ta Keo, Arcturus.   *She Ke, M. T. L.*

229   A.D. 892. *December.*

In the 1st year of the epoch King Fuh, the 11th moon, there was a comet in S. D. Tow and New.

Epoch King Fuh, A.D. 892–893 : 1st year, 892 : 11th moon, December.
S. D. Tow determined by ζ, τ, σ, ϕ, &c. in Sagittarius.
S. D. New determined by a, β, &c. Capricorni.   *She Ke, M. T. L.*

230   A.D. 893. *May 6.*

In the 2nd year of the same epoch, the 3rd moon, the heavens were for a long time covered with clouds. In the 4th moon, on the day Yih Yew, the clouds gradually opened, and a comet was seen in the evening in Shang Tae. It was about 100 cubits in length. It went to the east. It entered Tae Wei and swept Ta Keo. It entered Teen She. After 37 days it increased in length to about 200 cubits (*sic*), when the weather becoming cloudy it could no longer be seen.

Epoch King Fuh, 2nd year, 893 : 3rd moon, April : 4th moon, day Yih Yew, May 6th.
Shang Tae, ι, κ Ursæ Majoris.   Ta Keo, Arcturus.
Tae Wei, space within stars in Leo and Virgo.
Teen She, space bounded by Serpens.   *She Ke, M. T. L.*

Pingré has 895 for the year and June 25 for the day ; the Tables give the year 893, &c. as above.

231   A.D. 894. *February.*

In the 1st year of the epoch Keen Ning, the 1st moon, there was a comet in Shun Show.

Epoch Keen Ning, A.D. 894–897 : 1st year, 1st moon, 894, February.
Shun Show, one of the 12 kung, answering to our zodiacal sign Gemini or Cancer. It comprises the S. D. Tsing and Kwei, the stars composing which are in Gemini and Cancer.   *M. T. L.*

232   A.D. 905. *May 22.*

In the 2nd year of the epoch Teen Yew, the 4th moon, day Kea Shin, there was a comet in Ho Pih, Kwan, and Wan Chang. It was about 30 cubits in length. It entered Chung Tae and Hea Tae. In the 5th moon, on the day Yih Chow, in the evening, it was in the left angle of Heen Yuen, extending towards the west of Teen She. In the morning the luminous envelope had an exceedingly angry appearance. It extended across the heavens. On the day Ping Yin it was cloudy, and when, on the day Sin Wei, it ceased a little from raining, the comet was no longer visible.

Epoch Teen Yew, A.D. 904–905: 2nd year, 905: day Kea Shin, May 22: 5th moon, day Yih Chow, June 12; Ping Yin, June 13; Sin Wei, June 18.
Ho Pih, or Pih Ho, a, β, ρ, σ Geminorum.
Kwan, Corona Borealis.
Wan Chang, θ, ϕ, ν Ursæ Majoris.
Chung Tae, Hea Tae, stars in the feet of Ursa Major.
Heen Yuen, a and other stars in Leo and Leo Minor.
Teen She, space bounded by Serpens. *She Ke, M. T. L.*

Woo Tae, the Five Short Dynasties, A.D. 907–960.

How, or later Leang, A.D. 907–922.

233   A.D. 912. *May* 13.

In the reign of Tae Tsoo, the 2nd year of the epoch Keen Hwa, the 4th moon, day Jin Shin, a comet appeared in S. D. Chang. On the day Kea Seuh the comet was in Ling Tae.

Emperor Tae Tsoo, A.D. 907–912; epoch Kan Hwa, 911–912: 2nd year, 912: 4th moon, day Jin Shin, May 13; day Kea Seuh, May 15.
S. D. Chang determined by κ, λ, μ, &c. Hydræ.
Ling Tae, χ Leonis and small stars near. *She Ke, M. T. L.*

How, or later Tang, A.D. 923–935.

234   A.D. 928. *October* 14.

In the reign of Ming Tsung, the 3rd year of the epoch Teen Ching, the 10th moon, day Kang Woo, a comet appeared in the south-west. It was about 10 cubits in length. It pointed to the south-east. Its place was in the 5th degree of S. D. New. After three evenings it was no longer visible.

Emperor Ming Tsung and epoch Teen Ching, 926–929: 3rd year, 928: 10th moon, day Kang Woo, October 14.
S. D. New determined by a, β, &c. Capricorni. *She Ke, M. T. L.*

P

235   A.D. 936.   *October* 28.

In the reign of Fei Te, the 3rd year of the epoch Tsching Tae, the 9th moon, day Ke Chow, a comet appeared in S. D. Heu and Wei. It was about 1 cubit in length. It was very small. It passed the stars Teen Luy and Kuh.

> Emperor Fei Te, A.D. 934–935; epoch Tsching Tae, 3rd year, 936: 9th moon, day Yih Chow, October 28.
> S. D. Heu determined by β Aquarii and another.
> S. D. Wei determined by a Aquarii and θ, ϵ Pegasi.
> Teen Luy, ξ Aquarii, λ Capricorni, and others.
> Kuh, μ Capricorni.   *She Ke, M. T. L.*
> ' M. T. L.' has Mo Te for Fei Te.

How Tsin, A.D. 936–946.

236   A.D. 941.   *September* 18.

In the reign of Kaou Tsoo, the 6th year of the epoch Teen Fuh, the 9th moon, day Jin Tsze, a comet appeared in the west. It swept Teen She Yuen. It was about 10 cubits in length.

> Emperor Kaou Tsoo and epoch Teen Fuh, 936–944: 6th year, 941: 9th moon, day Jin Tsze, September 18.
> Teen She Yuen, space bounded by Serpens.   *M. T. L.*

237   A.D. 943.   *November* 5.

In the 8th year of the same epoch, 10th moon, day Kang Seuh, in the evening, a comet was seen in the east. It pointed to the west. The tail was 10 cubits in length. Its place was in the 9th degree of S. D. Keo.

> Teen Fuh, 8th year, 943: 10th moon, day Kang Seuh, November 5.
> S. D. Keo determined by a and ζ Virginis.   *She Ke, M. T. L.*
> ' M. T. L.' has 1 cubit in length.

How Chow, A.D. 951–960.

238   A.D. 956.   *March* 13.

In the reign of She Tsung, the 3rd year of the epoch Heen Tih, the 1st moon, day Jin Seuh, in the evening, there was a comet in S. D. Tsan. The tail pointed to the south-east.

> She Tsung, A.D. 954–959; epoch Heen Tih the same: 3rd year, 956: 1st moon, day Jin Seuh, March 13.
> S. D. Tsan determined by a, β, γ, &c. Orionis.   *She Ke, M. T. L.*

The later Sung Dynasty, a.d. 960–1279.

239   A.D. 975. *April.*

In the reign of Tae Tsoo, the 8th year of the epoch Kae Paou, the 3rd moon, a comet was seen in the east.

        Emperor Tae Tsoo, 960–975; epoch Kae Paou, 968–975: 8th year, 975: 3rd moon, April.

240   A.D. 975. *August 3.*

In the 6th moon of the same year, day Kea Tsze, a comet appeared in S. D. Lew. It was 40 cubits in length. In the morning it was seen in the east. It pointed to the south-west. It passed over Yu Kwei. It went on to the eastern part of S. D. Peih. Altogether it passed through 11 S. D. in 83 days, and then disappeared.

        Kae Yuen, 8th year, 975: 6th moon, day Kea Tsze, August 3.

        S. D. Lew determined by δ, ε, ζ, θ Hydræ.

        S. D. Peih determined by a Andromedæ and γ Pegasi.

        Yu Kwei, same as S. D. Kwei, determined by γ, δ, η, θ Cancri.   *M. T. L.*

241   A.D. 989. *August 13.*

In the reign of Tae Tsung, the 2nd year of the epoch Twan Kung, 6th moon, day Woo Tsze, there was a comet in the eastern part of S. D. Tsing, to the west of Tseih Shuwy. Its colour was a bluish white. Its luminous envelope gradually increased in length. In the morning it was seen for 10 days in the north-east, and to the north-west in the evening. It passed over Yew She Te. It was visible altogether for 30 days, after which it disappeared.

        Emperor Tae Tsung, A.D. 976–997; epoch Twan Kung, 988–989: 2nd year, 989: 1st moon, day Woo Tsze, August 13.

        S. D. Tsing determined by γ, ε, λ, μ, &c. Geminorum.

        S. D. Kang determined by ι, κ, λ, ϕ Virginis.

        Yew She Te, η, τ, ν Boötis.

        Tseih Shwuy, λ, μ Persei.   *M. T. L.*

242   A.D. 998. *February 23.*

In the reign of Ching Tsung, the 1st year of the epoch Han Ping, the 1st moon, day Kea Shin, there was a comet to the north of Ying Shih. Its luminous envelope was about 1 cubit in length. It passed on until the day Ting Yew, when it disappeared. It was altogether seen for 14 days.

        Emperor Ching Tsung, A.D. 998–1022; epoch Han Ping, 998–1003: 1st year, 998: 1st moon; day Kea Shin, February 23; day Ting Yew, March 8.

        Ying Shih, same as S. D. Shih, determined by a, β Pegasi, &c.   *M. T. L.*

243						A.D. 1003. *December 23.*

In the 6th year of the same epoch, the 11th moon, day Kea Yin, there was a comet in S. D. Tsing and Kwei. It was like a large cup. Its colour was a bluish white. Its luminous envelope was about 4 cubits in length. It entered Woo Choo Shih. It passed over Woo Chay and entered S. D. Tsan. It was visible altogether for about 30 days, after which it disappeared.

> Epoch Han Ping, 6th year, 1003 : 11th moon, day Kea Yin, December 23.
> S. D. Tsing determined by γ, ε, λ, μ, &c. Geminorum.
> Kwei determined by γ, δ, η, θ Cancri.
> Tsan determined by a, β, γ, δ, &c. Orionis.
> Woo Choo Shih, θ, ν, τ, &c. Geminorum. Woo Chay, a, β, &c. Aurigæ, &c.
>   *M. T. L.*

244						A.D. 1018. *August 4.*

In the 2nd year of the epoch Teen He, the 6th moon, day Sin Hae, a comet appeared in Pih Tow Kwei, to the north-east of the 2nd star. It was more than 3 cubits in length. It went to the north of the 1st star in Pih Tow. It passed near Teen Laou and over Wan Chang. Its length was then about 30 cubits. It passed through Tsze Wei, San Tae, and Heen Yuen. Its course was to the west until it arrived at Tseih Sing. It was visible altogether for 37 days, and then disappeared.

> Epoch Teen Hae, A.D. 1017–1021 : 2nd year, 1018 : 6th moon, day Sin Hae, August 4th.
> Pih Tow Kwei, the square in Ursa Major.
> Teen Laou, ω, &c. Ursæ Majoris. Wan Chang, θ, ν, φ, &c. Ursæ Majoris.
> San Tae, the stars in the feet of Ursa Major.
> Heen Yuen, Regulus and other stars in Leo and Leo Minor.
> Tseih Sing, the seven stars in S. D. Sing, determined by a, σ, τ, &c. Hydræ.
>   *M. T. L.*

245						A.D. 1035. *September 15.*

In the reign of Jin Tsung, the 2nd year of the epoch King Yew, the 8th moon, day Jin Seuh, in the evening, there was a comet in S. D. Chang and Yih. It was 7 cubits in length and $\frac{5}{10}$ths of a cubit in breadth. After 12 days it disappeared.

> Emperor Jin Tsung, A.D. 1023–1063 ; epoch King Yew, 1034–1037 : 2nd year, 1035 : day Jin Seuh, September 15.
> S. D. Chang determined by κ, λ, μ, &c. Hydræ.
> S. D. Yih determined by a, &c. Crateris.   *M. T. L.*

246						A.D. 1036. *January 15.*

In the 12th month of the same year, day Ke Wei, in the evening, a comet appeared in Wae Ping. It had a luminous envelope.

> 12th moon, day Ke Wei, 1036, January 15th.
> Wae Ping, a, β, ε and others in the band of Pisces.   *M. T. L.*

247   A.D. 1049. *March* 10.

In the 1st year of the epoch Hwang Yeu, the 2nd moon, day Ting Maou, a comet appeared in S. D. Heu. In the morning it was seen in the east. It pointed to the south-west. It passed through Tsze Wei and arrived at the S. D. Lew. It was visible for 114 days, and then disappeared.

Epoch Hwang Yeu, A.D. 1049–1053: 1st year, 1049: 2nd moon, day Ting Maou, March 10.

S. D. Heu determined by β Aquarii and another.

S. D. Lew determined by α, β, γ Arietis.

Tsze Wei, circle of perpetual apparition.   *M. T. L.*

248   A.D. 1056. *August.*

In the 1st year of the epoch Kea Yew, the 7th moon, a comet appeared in Tsze Wei. It passed through Tseih Sing. Its colour was white. It was about 10 cubits in length. It passed on until the 8th moon, day Kwei Hae, when it disappeared.

Epoch Kea Yew, A.D. 1056–1063: 1st year, 1056: 7th moon, August.

Tsze Wei, circle of perpetual apparition.

Tseih Sing, the seven stars. These appear to be the seven bright stars in Ursa Major.

249   A.D. 1066. *April* 2.

In the reign of Ying Tsung, the 3rd year of the epoch Che Ping, the 3rd moon, day Ke Wei, a comet appeared in Ying Shih. It was seen in the east in the morning, and was more than 7 cubits in length. It pointed to the south-west and to the S. D. Wei, extending to the stars Fun Moo. It gradually went afar off to the east. It approached the sun, and consequently could then not be seen. On the day Sin Sze it was again seen in the evening, to the north-west. It appeared like a star, without a bright envelope. It went to the east, increasing in size, and resembled a white vapour more than 3 cubits in breadth. It connected together Tsze Wei, Keih Sing, and S. D. Fang. The head and (the end of) the tail were obscured by clouds. It still went to the east. It passed Wan Chang and Pih Tow and crossed the S. D. Wei. On the day Jin Woo the star had regained its luminous envelope. The comet was then about 10 cubits in length and about 3 in breadth. It pointed to the north-east. It passed over Woo Chay, at which time the white vapour was divided into two branches. It crossed the heavens, passing through Pih Ho, Woo Choo How, Heen Yuen, Tae Wei, and Woo Te Tso, into Woo Choo How. It extended towards S. D. Keo, Kang, Te, and Fang. On the day Kwei Wei the comet was 15 cubits in length, and had round it a vapour resembling in form a Shing Ke (a kind of measure). Its course was thus from Ying Shih to S. D. Chang. In the north it altogether passed through 14 S. D. It was visible for 67 days, after which the star, the vapour, and the comet, all disappeared.

Q

Emperor Ying Tsung and epoch Che Ping, 1064–1067: 3rd year, 1066: 3rd moon, day Ke Wei, April 2: days, Sin Sze, April 24; Jin Woo, April 25; Kwei Wei, April 26.

S. D. Ying Shih, or Shih, determined by *a* Pegasi and stars near.
Wei determined by ε, μ, ν, &c. in Scorpio.
Keo determined by *a* Virginis and another.
Fang determined by *a*, δ, π, ρ in Scorpio.
Kang determined by ι, κ, λ, θ Virginis.
Te determined by *a*, β, γ, ν Libræ.
Chang determined by κ, λ, μ, &c. Hydræ.
Tsze Wei, circle of perpetual apparition.
Keih Sing, stars near the Pole.
Tae Wei, space between stars in Leo and Virgo.
Fun Moo, *a*, η, π and others in Aquarius.
Wan Chang, θ, ν, ϕ and others in Ursa Major.
Pih Tow, *a*, β, γ, δ, &c. in Ursa Major.
Woo Chay, *a*, β, &c. Aurigæ and β Tauri.
Pih Ho, *a*, β, &c. Geminorum.
Woo Choo How, θ, ι Geminorum, and also two groups of small stars between the head of Virgo and Coma Berenices. These must not be confounded together; the second Woo Choo How referred to in the text appearing to be the last-mentioned stars.
Heen Yuen, Regulus and stars in Leo and Leo Minor.
Woo Te Tso, β Leonis and other stars near. *M. T. L.*

It is singular that this very remarkable comet is not noticed in the 'She Ke.' In the 'Tung Keen Kang Moo' the account is as follows:—'In the reign of Yung Tsung, the 3rd year of the epoch Che Ping, a comet was seen in the west during the 3rd moon.' The 'Commentary' remarks, 'It resembled the planet Venus, and was 15 cubits in length. When it was in S. D. Peih it was like the moon.'
S. D. Peih determined by *a*, γ, δ, &c. Tauri.

This comet appears to have attracted much attention, and to have excited no little alarm in Europe, as we learn from contemporary writers that it was looked upon as a forerunner of various calamities: among others, the death of Harold and the subsequent conquest of England by William the Norman, is attributed by them to the influence of this comet. Zonares, the Greek historian, in his account of the reign of the Emperor Constantinus Ducas, describes it as having been as large as the full moon, and at first without a tail, on the appearance of which it diminished in size: thus corroborating the Chinese accounts, as given in 'M. T. L.' and the 'Tung Keen Kang Muh.'

250   A.D. 1095. *November* 17.

In the reign of Shin Tsung, the 8th year of the epoch He Ning, the 10th moon, day Yih Wei, a star appeared in the south-east, in the middle of the degrees of S. D. Chin. It was like the planet Saturn, of a bluish white. On the day Ping Shin it produced towards the north-west a luminous envelope, 3 cubits in length, pointing in a slanting direction to S. D. Chin. It thus resembled a comet. On the day Ting Yew the luminous envelope was 5 cubits in length. On the day Woo Seuh it was 7 cubits in length, pointing in a slanting direction towards Tso Hea. It went on until the day Ting Wei, when it entered the clouds and was no more seen.

Emperor Shin Tsung, 1068–1085; epoch He Ning, 1068–1077: 8th year, 1075: 10th moon, day Yih Wei, November 17; Ping Shin, November 18; Ting Yew, November 19; Woo Seuh, November 20; Ting Wei, November 29.

S. D. Chin determined by β and others in Corvus.

Tso Hea, η Corvi.

251   A.D. 1080. *August* 10.

In the 3rd year of the epoch Yuen Fung, the 7th moon, day Kwei Wei, a comet appeared in the north-west part of Tae Wei Yuen, to the south of Lang Wei. It resembled a white vapour, 10 cubits in length. It pointed in a slanting direction to the south-east. Its place was in the middle degrees of S. D. Chin. On the day Ping Seuh it went in a north-westerly direction. Its place was then in the middle degrees of S. D. Yih. On the day Woo Tsze it was 3 cubits in length, and went in a sloping direction across Lang Wei. On the day Kwei Maou (Kwei Sze) it entered Heen Yuen. On the day Ting Yew, the weather being thick, it could not be seen. On the day Kang Tsze it again appeared in the morning, in the middle degrees of S. D. Chang, until the day Woo Woo, when, having been visible altogether for 36 days, it disappeared, and was no more seen.

Epoch Yuen Fung, A.D. 1078–1085: 3rd year, 1080: 7th moon, day Kwei Wei, August 10; Ping Seuh, August 13; Woo Tsze, August 15; Kwei Sze (for Kwei Maou, see below), August 20; Ting Yew, August 24; Kang Sze, August 27; Woo Woo, September 14.

There is an obvious error in the original, the day Kwei Maou having been put for Kwei Sze. This is proved by summing up the days during which the comet was seen, which are said to have been 36. Reckoning Kwei Maou as one, they will amount to 96; whereas with Kwei Sze their number will be 36, as recorded above.

S. D. Chin determined by β and others in Corvus.

Yih determined by α and others in Crater.

Chang determined by κ, λ, μ and others in Hydra.

Tae Wei Yuen, space between stars in Leo and Virgo.

Lang Wei, stars in Coma Berenices.

Heen Yuen, Regulus and others in Leo and Leo Minor.

252   A.D. 1097. *October* 6.

In the reign of Che Tsung, the 4th year of the epoch Shaou Shing, the 8th moon, day Ke Yew, a comet appeared in the middle degrees of S. D. Te. It resembled the planet Saturn. It was like a bright white vapour, 3 cubits in length. It pointed in a slanting direction to the star Pa in Teen She Yuen. In the 9th moon, day Jin Tsze, the luminous envelope was 5 cubits in length. It entered Teen She Yuen. On the day Ke Wei it invaded Teen She Hwan. On the day Kang Shin it was near Te Tso in Teen She Yuen. On the day Woo Shin it disappeared, and was no more seen.

Emperor Che Tsung, A.D. 1086-1100; epoch Shaou Shing, 1094-1097: 4th year, 1097: 8th moon, day Ke Yew, October 6; days, Jin Tsze, October 9; Ke Wei, October 10; Kang Shin, November 6; Woo Shin, November 14.

S. D. Te determined by a, β, γ, ι Libræ.

Teen She Yuen, space bounded by Serpens.

Teen She Hwan, unascertained.

Pa, ε Serpentis.   Te Tso, a Herculis.   *M. T. L.*

253   A.D. 1106. *February* 10.

In the reign of Hwuy Tsung, the 5th year of the epoch Tsung Ning, the 1st moon, day Woo Seuh, a comet appeared in the west. It was like a great Pei Kow. The luminous envelope was scattered. It appeared like a broken-up star. It was 60 cubits in length and was 3 cubits in breadth. Its direction was to the north-east. It passed S. D. Kwei. It passed through S. D. Lew, Wei, Maou, and Peih. It then entered into the clouds and was no more seen.

Emperor Hwuy Tsung, A.D. 1101-1125; epoch Tsung Ning, 1102-1106: 5th year, 1106: 1st moon, day Woo Seuh, February 10.

S. D. Kwei determined by β, δ, ε Andromedæ and stars in Pisces.

Lew determined by a, β, γ Arietis.

Wei determined by the three stars in Musca.

Maou determined by the Pleiades.

Peih determined by a, γ, δ, ε, &c. Tauri.

Pei Kow is a kind of vessel or measure.   *M. T. L.*

This appears to have been a large meteor, as it seems to have been seen for a short time only.

254   A.D. 1110. *May* 29.

In the 4th year of the epoch Ta Kwan, the 5th moon, day Ting Wei, a comet appeared in S. D. Kwei and Lew. Its luminous envelope was 6 cubits in length. It went to the north and entered Tsze Wei Yuen. When in the north-west it entered the clouds and was no more seen.

Epoch Te Kwan, A.D. 1107-1110: 4th year, 1110: 5th moon, day Ting Wei. May 29th.

S. D. Kwei determined by β, δ, ϵ, &c. Andromedæ and stars in Pisces.

S. D. Lew determined by a, β, γ Arietis.

Tsze Wei Yuen, circle of perpetual apparition.   *M. T. L.*

255   A.D. 1126. *May* 20.

In the reign of Kin Tsung, the 1st year of the epoch Tsing Kang, the 6th moon, day Jin Seuh, a comet appeared in Tsze Wei Yuen.

The Commentary in the 'Tung Keen Keang Muh' adds, 'Its length was reckoned at 10 cubits. Its direction was to the north. It passed over Te Tso and swept Wan Chang.'

Emperor Kin Tsung and epoch Tsing Kang, A.D. 1126: 1st year, 1126: 6th moon, day Jin Seuh, May 20.

Tsze Wei Yuen, circle of perpetual apparition.

Te Tso, possibly Polaris, which is named Te. a Herculis has also the same appellation.

Wan Chang, ϕ, μ, ν Ursæ Majoris.   *M. T. L., Tung Keen.*

256   A.D. 1126. *December.*

In the intercalary 11th moon of the same year a comet was seen in the horizon.

A.D. 1126: intercalary 11th moon, December.   *M. T. L.*

257   A.D. 1131. *September.*

In the reign of Kaou Tsung, the 1st year of the epoch Shaou Hing, the 9th moon, a comet was seen.

Emperor Kaou Tsung, A.D. 1127–1162; epoch Shaou Hing, 1131–1162: 1st year, 1131: 9th moon, September.   *M. T. L.*

258   A.D. 1132. *January* 5.

In the 12th moon of the same year a comet was seen on the day Woo Yin.

12th moon, 1132, day Woo Yin, January 5.   *M. T. L.*

259   A.D. 1132. *August* 14.

In the 2nd year of the same epoch, the 8th moon, day Kea Yin, a comet was seen in S. D. Wei. In the 9th moon, day Kea Seuh, it disappeared.

Epoch Shaou Hing, 2nd year, 1132: 8th moon, day Kea Yin, August 14; day Kea Seuh, September 3.

S. D. Wei determined by the three stars in Musca.   *M. T. L.*

R

260   A.D. 1145. *April 26.*

In the 15th year of the same epoch, the 4th moon, day Woo Yin, a comet appeared in the degrees of the southern S. D. In about 50 days it disappeared. On the day Ping Shin it was seen in the degrees of S. D. Tsan.

15th year of epoch Shaou Hing, A.D. 1145: 4th moon, day Woo Yin, April 26; day Ping Shin, May 14.

S. D. Tsan determined by a, β, &c. Orionis.   *M. T. L.*

261   A.D. 1145. *June 4.*

In the 5th moon of the same year, day Ting Sze, a comet was seen . . . . . Its colour was a bluish white.

1145: 5th moon, day Ting Sze, June 4.   *M. T. L.*

262   A.D. 1147. *January 6.*

In the 16th year of the same epoch, 12th moon, day Woo Seuh, a comet appeared in the south-west of S. D. Wei.

16th year, 1146: 12th moon, day Woo Seuh, 1147, January 6.

S. D. Wei determined by a Aquarii and θ, ε Pegasi.   *M. T. L.*

263   A.D. 1147. *February 12.*

In the 17th year of the same epoch, the 1st moon, day Yih Hae, a comet appeared in the north-east, in the S. D. Neu. On the 2nd day of the 2nd moon it was no longer visible.

17th year, A.D. 1147: 1st moon, day Yih Hae, February 12; 2nd moon, 2nd day, March 7.

S. D. Neu determined by ε, μ, ν, &c. Aquarii.   *M. T. L.*

264   A.D. 1151. *August 21.*

In the 22nd year of the same epoch, 7th moon, day Ping Woo, a comet was seen in the north-east, in S. D. Tsing. On the day Ting Wei the star was like the planet Jupiter. Its luminous envelope was 1 cubit in length.

2nd year, A.D. 1151: 7th moon, day Ping Woo, August 21; day Ting Wei, August 22.

S. D. Tsing determined by γ, ε, λ, &c. Geminorum.   *M. T. L.*

265   A.D. 1222. *September 15.*

In the reign of Ning Tsung, the 15th year of the epoch Kea Ting, the 8th moon, day Kea Woo, a comet appeared in Yew She Te. Its luminous envelope was 30 cubits in length. Its body was small, like the planet Jupiter. It was seen for 2 months. It passed through S. D. Te, Fang, and Sing, and then disappeared.

Emperor Ning Tsung, A.D. 1195–1224; epoch Kea Ting, 1208–1224: 15th year, 1222: 8th moon, day Kea Woo, September 15.

S. D. Te determined by a, β, γ, ν Libræ.

Fang determined by β, δ, π, &c. in Scorpio.

Sin determined by Antares and others in Scorpio.

Yew She Te, η, τ, ν Boötis.

This is the last of the comets recorded in the 'Encyclopædia of Ma Twan Lin.' Those which follow are taken chiefly from the Supplement to that work and the 'She Ke.'

266   A.D. 1232. *October* 18.

In the reign of Le Tsung, the 5th year of the epoch Shaou Ting, the intercalary 9th moon, day Kang Seuh, a comet appeared in S. D. Keo.

Emperor Le Tsung, A.D. 1225–1264; epoch Shaou Ting, 1228–1233: 5th year, 1232: intercalary 9th moon, day Kang Seuh, October 18.

S. D. Keo determined by a and ζ Virginis.

267   A.D. 1240. *January* 31.

In the 4th year of the epoch Kea He, the 1st moon, day Sin Wei, a comet appeared in Ying Shih.

Epoch Kea He, 1237–1240: 4th year, 1240: 1st moon, day Sin Wei, Jan. 31.

S. D. Shih determined by a, β Pegasi, &c.

Ying Shih, a Pegasi.

268   A.D. 1240. *February* 23.

In the 1st moon of the same year, day Kea Woo, a comet passed over Yuh Lang, to the north-west of the second star.

1240: 1st moon, day Kea Woo, February 23.

Yuh Lang, a, β, &c. Cassiopeiæ.

This may possibly be a continuation of the account of the preceding comet.

269   A.D. 1264. *July* 26.

In the 5th year of the epoch King Ting, 7th moon, day Kea Seuh, at night, a comet appeared in S. D. Lew. Its tail extended across the heavens. On the day Ke Maou it passed into S. D. Kwei. In the 8th moon, day Sin Sze, it entered S. D. Tsing. On the day Woo Woo it could not be seen. On the day Kea Tsze it returned, and was seen in S. D. Tsan. On the day Sin Wei it was resolved into a reddish vapour.

Epoch King Ting, 1260–1264 : 7th moon, day Kea Seuh, July 26 ; day Ke Maou, July 31 : 8th moon, day Sin Sze, August 2 ; Woo Woo, September 8 ; Kea Tsze, September 14 ; Sin Wei, September 21.

S. D. Lew determined by δ, ε, &c. Hydræ.
Kwei determined by γ, δ, η, θ Cancri.
Tsing determined by γ, ε, λ, μ, &c. Geminorum.
Tsan determined by α, β, γ, &c. Orionis.

LEAOU, A MINOR DYNASTY, A.D. 916–1125.

270   A.D. 941. *August* 7.

In the reign of Tae Tsung, the 4th year of the epoch Hwuy Tung, the 8th moon, day Jin Shin, there was a comet near the star Tsin.

Emperor Tae Tsung, 927–947 ; epoch Hwuy Tung, 938–946, 4th year : 8th moon, day Jin Shin, 941, August 9.

Star Tsin, κ Herculis.

271   A.D. 1014. *February* 10.

In the reign of Shing Tsung, the 3rd year of the epoch Kae Tae, the 1st moon, day Yih Wei, a comet was seen in the west.

Emperor Shing Tsung, 983–1031 ; epoch Kae Tae, 1012–1021 : 3rd year, 1014 : 1st moon, day Yih Wei, February 10.

272   A.D. 1066. *April* 24.

In the reign of Taou Tsung, the 2nd year of the epoch Han Yung, the 3rd moon, day Jin Woo, a comet was seen in the east.

Emperor Taou Tsung, 1055–1100 ; epoch Han Yung, or Han Ning, 1065–1074 : 2nd year, 1066 : 3rd moon, day Jin Woo, April 24.

273   A.D. 1080. *January* 6.

In the 5th year of the epoch Tae Kang, the 12th moon, day Ping Woo, a comet passed over S. D. Wei.

Epoch Tae Kang, 1075–1084 : 5th year, 1079 : 12th moon, day Ping Woo, 1080, January 6.

S. D. Wei determined by ε, μ, ν, &c. in Scorpio.

274   A.D. 1097. *December* 6.

In the 3rd year of the epoch Show Lung, the 10th moon, day Ke Sze, a comet was seen in the west.

Epoch Show Lung, 1095–1110 : 3rd year, 1097 : 10th moon, day Ke Sze, December 6th.

KIN, A MINOR DYNASTY, A.D. 1118-1236.

275   A.D. 1133. *September 29.*

In the reign of Tae Tsung, 10th year of the epoch Teen Hwuy, 8th moon, day Sin Hae, a comet appeared in Wan Chang.

> Emperor Tae Tsung and epoch Teen Hwuy, 1124-1135: 10th year, 1133: 8th moon, day Sin Hae, September 29.
> Wan Chang, θ, ν, φ, &c. Ursæ Majoris.

276   A.D. 1226. *September 13.*

In the reign of Seuen Tsung, the 6th year of the epoch Hing Ting, the 8th moon, day Ke Maou, a comet appeared in S. D. Keo and Kang, between Yew Che Te and Chow Ting. It pointed towards Ta Keo. In the 1st year of the epoch Yuen Kwang, 9th moon, day Ting Wei, it disappeared.

> Emperor Seuen Tsung, 1217-1228; epoch Heen Ting, 1221-1226: 6th year, 1226: 8th moon, day Ke Maou, September 13: epoch Yuen Kwang, 1227-1228; 9th moon, day Ting Wei, September 12.
> S. D. Keo determined by a and ξ Virginis.
> S. D. Kang determined by ι, κ, λ, θ Virginis.
> Yew Che Te, η, τ, ν Boötis.
> Chow Ting, small stars in Coma Berenices.
> Ta Keo, Arcturus.

277   A.D. 1237. *September 21.*

In the reign of the Emperor Gae Tsung, the 1st year of the epoch Teen Hing, the 9th moon, day Ke Yew, a comet was seen in the east. It was about 10 cubits in length, twisted and bent like an elephant's tusk. It appeared in S. D. Keo and Chin. It went to the south. On the 12th day it was 20 cubits in length. On the 16th day it could not be seen, on account of the brightness of the moon. On the 27th day, in the 5th watch of the night, it reappeared, and was seen in the south-east. It was then about 40 cubits in length. On the 1st day of the 10th moon it began to fade. It was visible altogether for 48 days.

> Emperor Gae Tsung, 1229-1237; epoch Teen Hing, 1st year, 1237: 9th moon, day Ke Yew, September 21. 5th watch of night, 1 to 3 A.M.
> S. D. Keo determined by a and ξ Virginis.
> S. D. Chin determined by β, &c. Corvi.

Biot places this comet under 1232, October 17. According to the Tables, 1232 was the 4th year of the epoch Ching Ta. Biot's day is right for 1232, but not for 1237. No comet is mentioned in the 'She Ke' as having been seen in 1232. The 'Tung Keen' says a comet was seen in that year in Keo, but gives no particulars.

The above is from the Supplement to 'Ma Twan Lin.'

YUEN DYNASTY, A.D. 1280–1367.

The whole of the descriptions which follow are from the Supplement to
' M. T. L.' and the ' She Ke.'

278   A.D. 1264.   *July 26.*

In the reign of She Tsoo, the 1st year of the epoch Che Yuen, 7th moon, a comet
appeared in S. D. Kwei.   In the evening it was seen to the north-west.   It passed
through Shang Tae and swept Wan Chang in Tsze Wei, as well as Pih Tow.   In
the morning it was seen in the north-east.   It was visible altogether for about
40 days.

> Emperor She Tsoo, 1264–1294; epoch Che Yuen the same: 1st year, 1264.
>
> According to the Chinese Chronological Tables, the Tartar Emperor She Tsoo
> commenced his reign over China A.D. 1280, which was the 17th year of his epoch,
> Che Yuen.   Hence the 1st year was 1264.   His Tartar name was Hwuh Peih
> Lee: hence the Kublai of European writers.
>
> S. D. Kwei determined by γ, δ, η, θ Cancri.
> Tsze Wei, the circle of perpetual apparition.
> Shang Tae, ι, κ in fore-foot of Ursa Major.
> Pih Tow, α, β, &c. Ursæ Majoris.

> The account in the ' She Ke ' differs considerably, having some additional par-
> ticulars.   It is as follows:—
>
> In the reign of She Tsoo, the 1st year of the epoch Che Yuen, in the autumn,
> day Kea Seuh, a comet appeared in S. D. Kwei and Lew.   In the evening it was
> seen in the north-west.   Its brightness illuminated the heavens.   It measured 100
> cubits in length.   It passed through Shang Tae.   It swept Tsze Wei, Wan Chang,
> and Pih Tow.   In the morning it was seen in the north-east.   It was visible alto-
> gether for about 40 days.
>
> Emperor She Tsoo, as above; Che Yuen, 1st year, 1264: 7th moon, day Kea
> Seu, July 26.
> S. D. Kwei determined by γ, δ, η, θ Cancri.
> S. D. Lew determined by δ, ε and others in Hydra.
> For the remaining asterisms see above.

279   A.D. 1277.   *March 9.*

In the 14th year of the same epoch, 2nd moon, day Kwei Hae, a comet appeared
in the north-east.   It was about 4 cubits in length.

> Che Yuen, 14th year, 1277: 2nd moon, day Kwei Hae, March 9.

280   A.D. 1293. *November* 7.

In the 30th year of the same epoch, 10th moon, day Kang Yin, a comet entered Tsze Wei Yuen. Its course was towards Tow Kwei. Its luminous envelope was more than 1 cubit in length. It was visible for 1 moon and then disappeared.

> Che Yuen, 30th year, 1293 : 10th moon, day Kang Yin, November 7.
> Tsze Wei Yuen, circle of perpetual apparition.
> Tow Kwei, the square in the seven stars of Ursa Major. The Pole-star is sometimes called Tow Kwei.

> Biot has Pih Tow for Tow Kwei.

281   A.D. 1299. *June* 24.

In the reign of Ching Tsung, the 2nd year of epoch Ta Tih, the 12th moon, day Kea Seuh, a comet appeared beneath the stars Tsze and Sun.

> Emperor Ching Tsung, 1295–1307 ; epoch Ta Tih, 1297–1307 : 2nd year, 1298 : 12th moon, day Kea Seuh, 1299, January 24.
> Tsze, λ Columbæ.   Sun, θ, κ Columbæ.

282   A.D. 1301. *September* 16.

In the 5th year of the same epoch, 9th moon, day Yih Chow, from the 8th moon, day Kang Shin, a comet appeared in 24° 40′ of the S. D. Tsing. It was like the great star in Nan Ho. Its colour was white. Its length was 5 cubits. Its direction was towards the north-west. It afterwards passed to the south of Wan Chang and Tow Kwei. It swept Tae Yang. It also swept Teen Ke of Pih Tow, Tsze Wei Yuen, San Kung, and the stars in Kwan So. Its length was about 10 cubits. It passed into Teen She Yuen, to the east of the stars Pa and Shuh, and to the south of the stars Leang and Tsow, and above the star Sung. It was then a full cubit in length. It was altogether visible for 46 days, and then (on the day first mentioned) disappeared.

It is to be remarked, that the description of this comet commences with the day Yih Chow, being that of its disappearance. A few words have been added to make the description more intelligible. It is only slightly mentioned in the 'She Ke.'

> Epoch Ta Tih, 3rd year, 1301 : 5th moon, day Kang Shin, September 16 ; 9th moon, day Yih Chow, October 31.
> S. D. Tsing determined by γ, ε, λ, &c. Geminorum.
> Tsze Wei Yuen, circle of perpetual apparition.
> Teen She Yuen, space bounded by Serpens.
> Nan Ho, a, β, &c. Canis Minoris. The great star, Procyon.
> Wan Chang, θ, ν, φ, &c. Ursæ Majoris.
> Tow Kwei, the square in the seven stars of Ursa Major.
> Tae Yang, χ Ursæ Majoris.

Pih Tow, *a*, *β*, *γ*, &c. Ursæ Majoris.
Kwan So, Corona Borealis.
Pa, *ε* Serpentis.   Shuh, *a*, *λ* Serpentis.
Leang, *δ* Ophiuchi.   Tsoo, *ι* Ophiuchi.
Sung, *η* Ophiuchi.   Teen Ke, *γ* Ursæ Majoris.

283   A.D. 1304. *February 3.*

In the 8th year of the same epoch, the 3rd moon, day Yih Chow, from the day Kang Seuh of the preceding 12th moon a comet was seen. It was nearly a full cubit in length. It pointed towards the south-east. Its colour was white. Its place was in the 11th degree of S. D. Shih. It gradually increased to about a cubit in length, and then it pointed towards the north-west. It swept Tang Shay and entered Tsze Wei Yuen, and (on the day first mentioned) disappeared. It was visible altogether for 74 days.

As in the account of the preceding comet the day of disappearance is placed first. The following extract from the annals of the Yuen dynasty in the 'She Ke,' may be of service in explaining this rather ambiguous mode of expression. It relates to the same comet, and is to be found in the division Yuen She, section 4.

Ta Tih, 7th year, 12th moon, day Kang Seuh, a comet about a cubit in length was seen in the 11th degree of S. D. Shih. It entered Tsze Wei Yuen. In the 8th year, 3rd moon, day Yih Chow, the comet began to disappear. It was visible altogether for 74 days.

Epoch Ta Tih, 7th year, 1303: 12th moon, day Kang Seuh, 1304, Feb. 3; 8th year, 3rd moon, day Yih Chow, April 18, 1304.

S. D. Shih determined by *a*, *β*, &c. Pegasi.

Tsze Wei Yuen, circle of perpetual apparition.

Tang Shay, *π* Cygni and stars in Andromeda and Lacerta, 22 in number.

284   A.D. 1313. *April 13.*

In the reign of Jin Tsung, the 2nd year of the epoch Hwang King, 3rd moon, day Ting Wei, a comet appeared in the eastern part of S. D. Tsing.

Emperor Jin Tsung and epoch Hwang King, A.D. 1312–1320: 2nd year, 1313: 3rd moon, day Ting Wei, April 13.

S. D. Tsing determined by *γ*, *ε*, *λ*, *μ*, &c. Geminorum.

285   A.D. 1315. *November 28.*

In the 2nd year of the epoch Yen Yew, the 11th moon, day Ping Woo, a strange star appeared, which afterwards became a comet. It entered Tsze Wei Yuen. It passed through the S. D. from Chin to Peih, being 15 of those divisions. The next year, 2nd moon, day Kang Yin, it disappeared.

Epoch Yen Yew, 1314–1320: 2nd year, 1315: 11th moon, day Ping Woo, November 28: 3rd year, 2nd moon, day Kang Yin, 1316, March 12.

S. D. Peih determined by γ Pegasi and α Andromedæ.

S. D. Chin determined by β and others in Corvus.

286   A.D. 1337. *May* 4.

In the reign of Shun Te, the 3rd year of the epoch Che Yuen, in the summer, 4th moon, day Kea Seuh, there was a comet in Yuh Lang. It remained until the 7th moon, day Sin Yin, when it finished its course in Kwan So.

Shun Te, 1333–1367; epoch Che Yuen, 1335–1340: 3rd year, 1337: 4th moon, day Kea Seuh, May 4; 7th moon, day Sin Yin, July 31.

Yuh Lang, α, β, η and others in Cassiopeia.

Kwan So, Corona Borealis.

Biot considers this comet as the same as the next. It is, however, treated as a separate one both in 'Ma Twan Lin' and in the 'She Ke,' in which there is no intimation that the comet which follows, although on the same page, is in any way connected with it. It is, therefore, treated as a separate comet here.

287   A.D. 1337. *June* 26.

In the reign of Shun Te, the 3rd year of the epoch Che Yuen, the 5th moon, a comet was seen to the north-east. It resembled the great star in Teen Chuen. Its colour was white. It was about 1 cubit in length. The tail pointed to the south-west. Its place was estimated to be in the 5th degree of S. D. Maou. On the day Woo Shin its course was to the south-west. On the succeeding days it gradually increased in velocity. On the day Sin Wei, of the 6th moon, the luminous envelope had lengthened to about 2 cubits. On the day Ting Chow it swept Shang Ching. On the day Ke Maou the luminous envelope had increased still more in length, being then about 3 cubits. It entered Yuen Wei. On the day Jin Woo it swept Hwa Kae and the star Keang. On the day Yih Yew it swept the great star Kow Ching, and extended to Teen Hwang Ta Te. On the day Ping Seuh it passed through Sze Foo and crossed Keu Sin. On the day Kea Woo it left Yuen Wei. On the day Ting Yew it passed out of Tsze Wei Yuen. On the day Woo Seuh it entered Kwan So and swept Teen Ke. In the 7th moon, day Kang Tsze, it swept Ho Keen. On the day Kwei Maou it passed the stars Ching and Tsin and entered Teen She Yuen. On the day Ping Woo it swept Lee Sze. On the day Ke Yew the moon was so bright that the luminous envelope could scarcely be distinguished. The comet left Teen She Yuen and swept the star Leang. On the day Sin Yew the luminous envelope had greatly diminished in length. It was then in S. D. Fang, above the star Keen Pe, and directly west of the middle star of the asterism Fa. It was not easy to ascertain exactly the place of the comet after it had gradually gone to the south. It was visible altogether for 63 days. Its course was from S. D. Maou to S. D. Fang, making altogether 15 S. D. through which it passed, and afterwards disappeared.

T

The preceding account is from the Supplement to 'Ma Twan Lin,' and it must be observed that in the original, as I have it, an error occurs, the epoch there given being Che Ching instead of Che Yuen. That this is really an error is proved by the following account of the same comet, as it is given in the 'She Ke :'—

'In the 3rd year of the epoch Che Yuen, the 5th moon, day Ting Maou, a comet was seen in the north-east. It was like the great star in Teen Chuen. Its colour was white. It was about 1 cubit in length. The tail pointed to the south-west. It was altogether visible for 63 days. (Its course was) from S. D. Maou to S. D. Fang. It passed through 15 S. D. and then disappeared.'

The error is accordingly corrected in the text given, and does not occur in Biot. The day of the comet's first appearance (Ting Maou), which does not appear in 'M. T. L.,' is also given in this extract from the 'She Ke.' The comet appears to have been very carefully observed, and its course registered, almost day by day, until it went so far to the south as to render the observations difficult, and, consequently, uncertain. It must also be noticed, that the comet is described as passing through 15 S. D.; viz. from Maou (the Pleiades) to Fang (stars in Scorpio). Now as the greater number of the observations were made while the comet was within the circle of perpetual apparition, where the degrees are greatly contracted, such a circumstance could easily occur.

Epoch Che Yuen, 3rd year, 1337 : 5th moon, day Ting Maou, June 26 : 6th moon, day Sin Wei, June 30 ; Ting Chow, July 6 ; Ke Maou, July 8 ; Jin Woo, July 11 ; Yih Yew, July 14 ; Ping Seuh, July 15 ; Kea Woo, July 23 ; Ting Yew, July 26 ; Woo Seuh, July 27 : 7th moon, Kang Tsze, July 29 ; Kwei Maou, August 1 ; Ping Woo, August 4 ; Ke Yew, August 7 ; Sin Yew, August 19.

S. D. Maou determined by the Pleiades.
S. D. Fang determined by β, δ, π, &c. in Scorpio.
Teen She Yuen, space bounded by Serpens.
Teen Chuen, α, β, δ, &c. Persei.   The great star, α Persei.
Shang Ching, A 579 Camelopardalis (Reeves).
Yuen Wei, stars in Draco.
Hwa Kae, stars in Camelopardalis.   Keang, unascertained.
Kow Ching, α Ursæ Majoris.
Teen Hwang Ta Te, Polaris.
Sze Foo, four small stars near the Pole.
Keu Sin, unascertained.
Kwan So, Corona Borealis.
Teen Ke, θ and other small stars in Hercules.
Ho Keen, γ Herculis.   Ching, γ Serpentis.   Tsin, χ Herculis.
Lee Sze, λ Ophiuchi and other stars near.
Leang, δ Ophiuchi.
Keen Pe, ν in Scorpio.
Fa, ε, ψ, o Libræ (Reeves).   Stars in Scorpio (Noel).

288   A.D. 1340. *March* 24.

In the 6th year of the same epoch, 2nd moon, day Ke Yew, a comet appeared resembling the great star in S. D. Fang. Its colour was white. In appearance it resembled a mass of the refuse of silk. Its length was about half a cubit. The tail pointed to the south-west. Its place was in the seventh degree of S. D. Fang. It went slowly to the north-west until the 3rd moon, day Kang Shin. It was altogether visible for 32 days.

Che Yuen, 6th year, 1340: 2nd moon, day Ke Yew, March 24; 3rd moon, day Kang Shin, April 24.

S. D. Fang determined by β, δ, π in Scorpio.

The great star in Fang, β in Scorpio. Possibly Antares is really the star meant.

289   A.D. 1351. *November* 24.

In the 11th year of the epoch Che Ching, on the day Sin Hae, a comet was seen in S. D. Kwei. On the day Kwei Chow it was seen in S. D. Lew. On the day Kea Yin it was in S. D. Wei. On the day Yih Maou it was still in that division. On the day Ping Shin it was seen in S. D. Maou. On the day Ting Sze it was seen in S. D. Peih.

Epoch Che Ching, 1341–1367: 11th year, 1351: 11th moon, day Sin Hae, November 24; Kwei Chow, November 26; Kea Yin, November 27; Yih Maou, November 28; Ping Shin, November 29; Ting Sze, November 30.

S. D. Kwei determined by a and others in Andromeda and Pisces.

Lew determined by a, β, γ Arietis.

Wei determined by the three stars in Musca.

Maou determined by the Pleiades.

Peih determined by a, γ, δ, &c. Tauri.

290   A.D. 1356. *September* 21.

In the 16th year of the same epoch, the 8th moon, day Kea Seuh, a comet was seen precisely in the east. It appeared in Heen Yuen, in the angle to the left of the great star in that asterism. Its colour was a bluish white; the tail pointed to the south-west. Its length was about 1 cubit. It was in $17°\frac{1}{10}$ of the S. D. Chang. In the 10th moon, day Woo Woo, it disappeared. It was traced to the north-west for about 40 days.

Epoch Che Ching, 16th year, 1356: 8th moon, day Kea Seuh, September 1.

Heen Yuen, Regulus and γ, ε, η, λ and others in Leo and Leo Minor.

The great star in Heen Yuen, Regulus.

291   A.D. 1360. *March* 12.

In the 20th year of the same epoch, 3rd moon, day Woo Tsze, there was a comet in the north-east.

Che Ching, 20th year, 1360: 3rd moon, day Woo Tsze, March 12.

292   A.D. 1362. *March 5.*

In the 22nd year of the same epoch, 2nd moon, day Yih Yew, a comet was seen. Its luminous envelope was about a cubit in length. Its colour was a bluish white. Its place was in 7° 20′ of S. D. Wei. On the day Ting Yew the comet passed near the western star of Le Kung. At the end of the 2nd moon the luminous envelope was about 20 cubits in length. In the 3rd moon, day Woo Shin, the comet could not be seen as a star, but only as a white vapour of a curved form, extending across the heavens and pointing to the west. It swept Ta Keo. On the day Jin Sze the comet passed before Tae Yang; it had then the appearance of a star without a tail. In form it resembled a great wine-cup. The colour was white, like the obscure twilight. Its place was in the 6th degree of S. D. Maou. On the day Woo Woo it began to disappear.

The account of this comet in the 'She Ke' commences thus:—'On the day Yih Yew a comet was seen in S. D. Wei. Its luminous envelope was about *ten* cubits in length.' The remainder is nearly the same as in 'M. T. L.,' the difference being merely verbal.

Che Ching, 22nd year, 1362: 2nd moon, day Yih Yew, March 5; Ting Yew, March 17; Woo Shin, March 28; Sin Tsze, April 1; Woo Woo, April 7.
S. D. Wei determined by *a* Aquarii and θ, ε Pegasi.
S. D. Maou determined by the Pleiades.
Le Kung, three groups of two stars each in Pegasus: they are λ μ, η o, ν τ.
Ta Keo, Arcturus.   Tae Yang, χ Ursæ Majoris.

293   A.D. 1362. *June 29.*

In the same year, the 6th moon, day Sin Sze, a comet was seen in Tsze Wei Yuen. Its place was in $2°\frac{90}{100}$ of S. D. New. Its colour was white. Its luminous envelope was about a cubit in length, pointing to the south-east. Its course was to the south-west. On the day Woo Tsze the luminous envelope of the comet swept Shang Tsae. In the 7th moon, day Yih Maou, it began to disappear.

Che Ching, 22nd year, 1362: 6th moon, day Sin Sze, June 29; day Woo Tsze, July 6: 7th moon, day Yih Maou, August 2.
S. D. New determined by *a*, β, &c. Capricorni.
Tsze Wei Yuen, circle of perpetual apparition.
Shang Tsae, θ Draconis.

294   A.D. 1363. *March 16.*

In the 23rd year of the same epoch, 3rd moon, day Sin Chow, the 1st day of the moon, a comet was seen in the east. It was visible during that moon, and then disappeared.

Che Ching, 23rd year, 1363: 3rd moon, day Sin Chow, March 16.

295   A.D. 1366. *October 25.*

In the 26th year of the same epoch, 9th moon, day Kang Sze, a comet was seen in Tsze Wei Yuen, near the star Kwan in Pih Tow. Its colour resembled that of a handful of meal. It appeared nearly as large as a Tow measure. Its course was to the south-east, and it passed near to the star Teen Keae. On the day Sin Chow the place of the comet was in $18°\frac{50}{100}$ of S. D. Wei. On the day Sin Yin it was in $2°\frac{50}{100}$ of S. D. New. On the day Kwei Maou the comet was in $9°\frac{90}{100}$ of S. D. New. On the day Kea Shin it was in $0°\frac{80}{100}$ of S. D. Heu. On the day Yih Sze the comet appeared in Tsze Wei Yuen, between the stars Kwan and Yuh Kang in Pih Tow. It was then in S. D. Chin. It went to the south-east and passed over Teen Kae. It traversed Tsan Tae, Leen Taou, and S. D. Heu, to the western star of Luy Peih Chin, when it began to disappear.

Such is a nearly literal translation of the account of this comet in the Supplement to 'Ma Twan Lin,' and it is not at all surprising that Biot should make the following remark, 'La marche indiquée pour cette comète est très singulière,' as nothing can be more inconsistent than that a comet, after a long course from Ursa Major to Aquarius, where it was observed on Oct. 29, should on the very next day, Oct. 30, be found once more in Ursa Major, in the same place whence it started, and again take its course southward, in the same direction as at first. But if the narrative in the original be carefully examined, it will be found to divide readily into two distinct portions; the one giving the course of the comet through the S. D., and the other that through some of the asterisms in that course. All that is required is to read the account according to this view, and a consistent narrative will be the result. Let, then, the concluding observations be read thus,—'On the day Yih Sze (October 30) the comet (after having been first seen in Tsze Wei Yuen, between the stars Kwan and Yuh Kang in Pih Tow, at which time it was in S. D. Chin, then going to the east, passing near Teen Kae, and traversing Tsan Tae, Leen Taou, and S. D. Heu), arrived at the western star of Luy Peih Chin, where it disappeared.' In corroboration of this view it may also be remarked, that the asterisms mentioned in this second portion will all be found in the path of the comet through the S. D. mentioned, supposing them to be carried to the Pole; and thus the whole account becomes perfectly consistent. It must also be remarked, that in these Chinese accounts of comets there are several examples of the latest observation being that first mentioned. Thus, in the comets of 1301, September 16, and 1315, November 28, the day of the disappearance is placed first, as in the second portion of the preceding narrative.

Che Ching, 26th year, 1366: 9th moon, day Kang Sze, October 25; Sin Chow, October 26; Jin Yin, October 27; Kwei Maou, October 28; Kea Shin, October 29; Yih Sze, October 30.

S. D. Wei determined by ε, μ, ν, &c. Aquarii,
       New determined by ε, μ, ν, &c. in Scorpio,
       Heu determined by β Aquarii, &c.
       Chin determined by β Corvi and others,
Tsze Wei Yuen, circle of perpetual apparition,

Kwan, δ Ursæ Majoris.
Yuh Kang, ε Ursæ Majoris.
Pih Tow, α, β, &c. Ursæ Majoris.
Teen Kae, β, γ Draconis.
Tsan Tae, β, δ, &c. Lyræ.
Leen Taou, η, θ Lyræ.
Luy Peih Chin, small stars in Aquarius and Pisces.

MING DYNASTY, A.D. 1368–1644.

296   A.D. 1368. *February 7.*

In the reign of Tae Tsoo, the 1st year of the epoch Hung Woo, 1st moon, day Kang Yin, a comet was seen in S. D. Maou and Peih.

> Emperor Tae Tsoo and epoch Hung Woo, 1368–1398: 1st year, 1368: 1st moon, day Kang Yin, February 7.
> S. D. Maou determined by the Pleiades.
> S. D. Peih determined by α, γ, δ, ε, &c. Tauri.

297   A.D. 1368. *April 8.*

In the 3rd moon of the same year, day Sin Maou, a comet appeared in S. D. Maou, to the north, between Ta Ling and Teen Chuen. It was about 8 cubits in length, and pointed towards Wan Chang. It came near Woo Chay. In the 4th moon, day Ke Yew, it disappeared to the north of Woo Chay.

> 1368: 3rd moon, day Sin Maou, April 8; Ke Yew, April 26.
> S. D. Maou determined by the Pleiades.
> Ta Ling, β, &c. Persei.
> Woo Chay, α, β, &c. Aurigæ and β Tauri.
> Wan Chang, θ, ν, φ Ursæ Majoris.
> Teen Chuen, α, γ, δ, &c. Persei.

> This was possibly the same as the preceding comet.

298   A.D. 1373. *May.*

In the 6th year of the same epoch, 4th moon, three comets entered Tsze Wei Yuen.

> Hung Woo, 6th year, 1370: 4th moon, May.
> Tsze Wei Yuen, circle of perpetual apparition.

299   A.D. 1391. *May 23.*

In the 24th year of the same epoch, 4th moon, day Ping Tsze, there were two comets. One entered Tsze Wei Yuen by the Chung Ho gate. It passed near Teen Chwang. The other passed near Luh Kea, and swept Woo Te Nuy Tso.

Hung Woo, 24th year, 1391 : 4th moon, day Ping Tsze, May 13.
Tsze Wei Yuen, circle of perpetual apparition.
Chung Ho Mun, space between *a* and *ι* Draconis.
Teen Chwang, small stars near *θ* Draconis.
Luh Kea, small stars in Camelopardalis.
Woo Te Nuy Tsoo, small stars near Polaris.

300   A.D. 1407. *December* 14.

In the reign of Ching Tsoo, 5th year of epoch Yung Lo, 11th moon, day Ping Yin, a comet was seen.

Emperor Ching Tsoo and epoch Yung Lo, 1403–1424 : 5th year, 1407 : day Ping Yin, December 14.

301   A.D. 1431. *May* 15.

In the reign of Seuen Tsung, the 6th year of the epoch Seuen Tih, 4th moon, day Woo Seuh, there was a comet in the eastern part of S. D. Tsing. It was about 5 cubits in length.

Emperor Seuen Tsung and epoch Seuen Tih, 1426–1435 : 6th year, 1431 : 4th moon, day Woo Seuh, May 15.
S. D. Tsing determined by *γ*, *ε*, *λ*, &c. Geminorum.

Biot makes the date of this Kang Seuh May 27, which is also correct as to the day, it being a subsequent date.

302   A.D. 1432. *February* 3.

In the 7th year of the same epoch, the 1st moon, day Jin Seuh, a comet appeared in the east. It was about 10 cubits in length : the tail swept Teen Tsin. It went to the south-east. In the 10th moon it began to disappear.

Biot has, 'After 10 days it began to disappear,' which is the most probable reading. It is not in 'M. T. L.'

Seuen Tih, 7th year, 1432 : 1st moon, day Jin Seuh, February 3 ; 10th moon, November.
Teen Tsin, *a*, *γ*, *δ*, *ε* and others in Cygnus.

303   A.D. 1432. *February* 29 or *October* 26.

In the same moon, on the day Woo Tsze, another comet appeared in the west. After 17 days it disappeared.

It is not clear whether this refers to the 1st or to the 10th moon. If the 1st, then Woo Tsze will be February 29 ; if the 10th, October 26.

304   A.D. 1433. *September* 15.

In the 8th year of the same epoch, the intercalary 8th moon, day Jin Sze, a comet appeared in Teen Tsang. It was more than 10 cubits in length. On the day Ke Sze it entered Kwan So and swept Tseih Kung. On the day Ke Maou it again entered Teen She Yuen and swept the star Tsin. It was visible for 24 days, and then disappeared.

      Seuen Tih, 8th year, 1433: intercalary 8th moon, day Jin Tsze, September 15; Ke Sze, October 2; Ke Maou, October 12.

      Teen Tsang, θ, ι, κ Boötis.

      Kwan So, Corona Borealis.   Tsin, α Herculis.

      Tseih Kung, δ, μ and others in hand of Boötes.

      Teen She Yuen, space bounded by Serpens.

305   A.D. 1439. *March* 25.

In the reign of Ying Tsung, the 4th year of epoch Ching Tung, the intercalary 2nd moon, day Ke Chow, a comet was seen in S. D. Chang. It was large, and like a ball. On the day Ting Yew it was about 50 cubits in length. It went to the west. It swept Tsew Ke. It then went to the north and passed into S. D. Kwei.

      Emperor Ying Tsung and epoch Ching Tung, 1436–1439: Ching Tung, 4th year, 1439; intercalary 2nd moon, day Ke Chow, March 25: Ting Yew, April 2.

      S. D. Chang determined by δ, κ, λ, μ Hydræ.

      S. D. Kwei determined by γ, δ, η, θ Cancri.

      Tsew Ke, ξ, ψ, ω Leonis and κ, ξ Cancri.

306   A.D. 1439. *July* 12.

In the 6th moon of the same year, day Woo Yin, a comet was seen in S. D. Peih, near the asterism so called. It was about 10 cubits in length. It pointed towards the south-west. It was visible altogether for 55 days, and then disappeared.

      1439: 6th moon, day Woo Yin, July 12.

      S. D. Peih determined by γ Pegasi and α Andromedæ.

307   A.D. 1444. *August* 6.

In the 9th year of the same epoch, the 7th moon, day Kang Woo, a comet was seen in Tae Wei Yuen, to the east. It was more than 10 cubits in length. It gradually increased in length until the intercalary 7th moon, day Ke Maou, when it entered S. D. Keo and disappeared.

      Ching Tung, 9th year, 1444: 7th moon, day Kang Woo, August 6; Ke Maou, August 15.

      S. D. Keo determined by α and ζ Virginis.

      Tae Wei Yuen, space within stars in Leo and Virgo.

308   A.D. 1449. *December* 20.

In the 14th year of the same epoch, 12th moon, day Jin Tsze, a comet was seen in Teen She Yuen, near to She Low. It passed through the degrees of S. D. Wei. It was 2 cubits in length. It was seen until the day Yih Hae, when it disappeared.

      Ching Ting, 14th year, 1449 : 12th moon, day Jin Tsze, December 20 ; day Yih Hae, 1450, January 12.

      S. D. Wei determined by ε, μ, ν, &c. in Scorpio.

      She Low, μ Ophiuchi.

      Teen She Yuen, space bounded by Serpens.

309   A.D. 1450. *January* 19.

In the reign of King Te, the 1st year of the epoch King Tae, the 1st moon, day Jin Woo, a comet appeared just without the boundary of Teen She Yuen. It swept Teen Ke.

      Emperor King Te and epoch King Tae, 1450–1456 : 1st year, 1450 : 1st moon, day Jin Woo, January 19.

      Teen She Yuen, space bounded by Serpens.

      Teen Ke, small stars near θ Herculis.

      This is most likely the same comet as the preceding one.

310   A.D. 1452. *March* 21.

In the 3rd year of the same epoch, 3rd moon, day Kea Woo, the 1st day of the moon, there was a comet in S. D. Peih.

      King Tae, 3rd year, 1452 : 3rd moon, day Kea Woo, March 21.

      S. D. Peih determined by α, γ, δ, ε, &c. Tauri.

      Biot makes this 2nd moon March 5. March 21 is correct for the day Kea Woo. March 5, in 1452, was Woo Yin.

311   A.D. 1456. *May* 27.

In the 7th year of the same epoch, the 4th moon, day Jin Seuh, a comet was seen to the north-east, in S. D. Wei. It was 2 cubits in length, and pointed towards the south-west. In the 5th moon, day Kwei Yew, it gradually lengthened to about 10 cubits. On the day Woo Tsze it was seen to the north-west, in S. D. Lew. It was then about 9 cubits in length. It swept over the stars Heen Yuen. On the day Kea Woo it was seen in S. D. Chang. It was then about 7 cubits in length. It swept the north of Tae Wei. It went to the south-west. In the 6th moon, day Jin Yin, it entered Tae Wei Yuen. It was then about 1 cubit in length.

      King Tae, 7th year, 1456 : 4th moon, day Jin Seuh, May 27 ; 5th moon, day Kwei Yew, June 7 ; 6th moon, day Jin Yin, July 6.

      S. D. Wei determined by the three stars in Musca.

                                                x

S. D. Lew determined by δ, ε, ζ, θ Hydræ.
  Chang determined by κ, λ, μ, &c. Hydræ.
 Heen Yuen, Regulus and stars in Leo and Leo Minor.
 Tae Wei Yuen, space between stars in Leo and Virgo.

312   A.D. 1457. *January* 14.

In the 12th moon of the same year, day Kea Yin, another comet was seen in S. D.
Peih. It was half a cubit in length. It went to the south-east. It gradually lengthened
until the day Kwei Wei, when it disappeared.

  King Tae, 7th year, 1456: 12th moon, day Kea Yin, 1457, January 14; Kwei
Hae, January 23.
  S. D. Peih determined by α, γ, δ, ε, &c. Tauri.

313   A.D. 1457. *June* 15.

In the reign of Ying Tsung, the 1st year of the epoch Teen Shun, the 5th moon,
day Ping Seuh, a comet was seen in S. D. Wei. It was like the star Chaou Yaou. It
went to the east. Its luminous envelope was half a cubit in length, pointing to the
south-west. In the 6th moon, day Kwei Sze, the 1st day of the moon, it was seen in
S. D. Shih. It was then about 10 cubits in length: the tail extended to the east of
S. D. Peih, and was near Teen Ta Tseang Keun, the 3rd star in Keuen She, S. D. Tsing,
and the 2nd southern star in Schwuy Wei.

  Emperor Ying Tsung, 1436–1464. This Emperor was taken prisoner by the
Tartars in 1450 and restored in 1457, when he adopted the epoch Teen Shun,
1457–1464: 1st year, 1457: 5th moon, day Ping Seuh, June 15; day Kwei Sze,
June 22.
  S. D. Wei determined by α Aquarii and θ, ε Pegasi.
   Shih determined by α, β Pegasi and others.
   Peih determined by γ Pegasi and α Andromedæ.
   Tsing determined by δ, ε, λ, μ, &c. Geminorum.
 Chaou Yaou, β Boötis. Keuen She, ν Persei.
 Teen Ta Tseang Keun, γ and others in Andromeda and Triangulum.
 Shwuy Wei, ζ, θ, o, π Canis Minoris.

314   A.D. 1457. *October* 26.

In the 10th moon of the same year, day Ke Hae, a comet was seen in S. D. Keo.
It was about half a cubit in length, pointing to the north. It passed near the northern
star of Keo and the eastern star of Ping Taou.

  1457: 10th moon, day Ke Hae, October 26.
  S. D. Keo determined by Spica and another in Virgo.
  Northern star, ζ Virginis.
  Ping Taou, θ and another in Virgo.

315   A.D. 1461. *August 5.*

In the 5th year of the same epoch, 6th moon, day Woo Seuh, a comet was seen in the east. It pointed to the south-west. It entered S. D. Tsing. In the 7th moon, day Ping Yin, it began to disappear.

> Teen Shun, 5th year, 1461 : 6th moon, day Woo Seuh, August 5; day Ping Yin, September 2.
> S. D. Tsing determined by γ, ε, λ, &c. Geminorum.

316   A.D. 1465. *March.*

In the reign of Heen Tsung, 1st year of the epoch Ching Hwa, 2nd moon, a comet was seen. In the 3rd moon it was again seen, in the north-west. It was about 30 cubits in length. It was visible during the 3rd moon, and then disappeared.

> Emperor Heen Tsung and epoch Ching Hwa, 1465–1467 : 1st year, 1465 : 2nd moon, March; 3rd moon, April.

317   A.D. 1468. *September* 18.

In the 4th year of the same epoch, the 9th moon, day Ke Wei, there was a star seen in the 5th degree of S. D. Sing. For 5 days it went to the north-east. Its luminous envelope was about 30 cubits in length : the tail pointed to the south-west. It changed into a comet. It was afterwards seen in the morning, in the east. In the evening it was seen in the south of S. D. Shih. It passed through San Kang, Pih Tow, Yaou Kwang, and Tseih Kung. It turned and entered Teen She Yuen. It left Yuen and gradually decreased. It passed over the first star to the west of Teen Ping. In the 11th moon, day Kang Shin, it disappeared.

> Ching Hwa, 4th year, 1468 : 9th moon, day Ke Wei, September 18 ; Kang Shin, December 8.
> S. D. Sing determined by α and others in Hydra.
> S. D. Shih determined by α, β Pegasi, &c.
> Yaou Kwang, η Ursæ Majoris.
> Tseih Kung, δ, μ and others in Boötes.
> Teen She Yuen, space bounded by Serpens.
> Teen Ping, unascertained.

318   A.D. 1472. *January* 16.

In the 7th year of the same epoch, 12th moon, on the day Kea Seuh, a comet was seen in Teen Teen. It pointed towards the west. It suddenly went to the north. It passed through Yew She Te. It swept Shang Tseang in Tae Wei Yuen, and also Hing Chin, Tae Tsze, and Tsung Kwan. The tail pointed directly to the west. It swept across Tae Wei Yuen and Lang Wei. On the day Ke Maou the luminous envelope had lengthened greatly. It extended from east to west across the heavens. It went northwards about 28 degrees. It passed near Teen Tsang and swept Pih Tow, San Kung, and Tae Yang. It entered Tsze Wei Yuen, and is said to have been seen in full day-light. It passed near to the stars Te Hwang, Kwei in Pih Tow, Shoo Tsze, How Kung, Kow Shin, Teen Choo, San Sze, Leen Taou, Chung Tae, Teen Hwang, Ta Te, Shang Wei, Ko Taou, Wan Chang, and Shang Tae. On the day Yih Yew it went to the south, and passed through Teen Ho, Teen Yen, Wae Ping, and Teen Yuen. In the 1st moon of the 8th year, on the day Ping Woo, it went towards the group Wae Ping, in S. D. Kwei. It gradually faded, and it was some time before it finally disappeared.

Ching Hwa, 7th year, 1471 : 12th moon, day Kea Seuh, 1472, January 16 ; Ke Maou, January 21 ; Yih Yew, January 27 : 8th year, 1st moon, day Ping Woo, 1472, February 17.

S. D. Lew determined by a, β, γ Arietis.
S. D. Kwei determined by β, δ, ϵ, &c. Andromedæ, and stars in Pisces.
Tae Wei Yuen, space between stars in Leo and Virgo.
Tsae Wei Yuen, circle of perpetual apparition.
Yew She Te, η, ν, τ Boötis.   Teen Teen, σ, τ Virginis.   Tae Tsze, ϵ Leonis.
Hing Chin, star in Coma Berenices.   Tung Kwan, 2567 Leonis.
Teen Tseang, ν, θ, κ Boötis.
Pih Tow, the seven bright stars in Ursa Major.
San Kung, the three stars in the head of Asterion.
Tae Yang, χ Ursæ Majoris.
Teen Hwang Ta Te, Polaris.   Star Te Hwang, β Ursæ Minoris.
Kwei in Pih Tow, the square in Ursa Major.
Shoo Tsze, A 3233 Ursæ Minoris.   How Kung, b 3162 Ursæ Minoris.
Kow Chin, ζ Ursæ Minoris.   Teen Choo, a Ursæ Majoris.
San Tsze, ϕ, σ, &c. Ursæ Majoris.
Teen Laou, ω and small stars in Ursa Major.
Chung Tae, λ, μ, Shang Tae, ι, κ Ursæ Majoris.
Wan Chang, θ, υ, ϕ Ursæ Majoris.
Shang Wei, star in Camelopardalis; also one in Cepheus.
Ko Taou, ν, ξ, o, π Cassiopeiæ.
Teen Ho, e, b, o, z Arietis.   Tsin Yin, δ, ζ Arietis.
Wae Ping, a, δ, ϵ, &c. in Pisces.   Teen Yuen, stars in Cœtus.

319   A.D. 1490. *December* 31.

In the reign of Heaou Tsung, the 3rd year of the epoch Hung Che, the 11th moon, day Woo Seuh, a comet was seen in the south of Teen Tsin. Its tail pointed to the north-east. It passed over the star Jin. It passed through Choo Kew. In the 12th moon, day Woo Shin, being the 1st day of the moon, it entered Ying Shih. On the day Kang Shin it passed into Teen Tsang.

Emperor Heaou Tsung and epoch Hung Che, 1488–1505; 3rd year, 1490 : 12th moon, day Woo Seuh, December 31 : days, Woo Shin, 1491, January 10 : Kang Shin, January 22.

There appears to be an error in the original in the moon. The Supplement to 'M. T. L.' has the 12th moon instead of the 11th, which seems to be correct. Pingré, after Gaubil, has the 12th moon; and Biot's computations agree with that moon, but are not consistent with the 11th moon. The 12th moon has, therefore, been employed instead of the 11th in the preceding computations of the dates.

Teen Tsin, a, β, γ, δ and others in Cygnus.

Jin Sing, e, f, g Pegasi.

Ying Shih, or S. D. Shih, determined by a Pegasi and others.

Teen Tsang, ι, η, θ, τ Ceti.

320   A.D. 1500. *May 8.*

In the 13th year of the same epoch, 4th moon, day Kea Woo, a comet was seen in Luy Peih Chin. It entered the space between S. D. Shih and Peih. It gradually lengthened, until it was about 3 cubits in length. It pointed towards Le Kung and swept Tsaou Foo. It passed Tae Wei Yuen. It gradually lessened, and entering Tsze Wei Yuen it approached near to New She. It passed through Shang Shoo. In the 6th moon, day Ting Yew, it disappeared.

Hung Che, 13th year, 1500: 4th moon, day Kea Woo, May 8; 6th moon, day Ting Yew, July 10.

S. D. Shih determined by a, β Pegasi, &c.

S. D. Peih determined by γ Pegasi and a Andromedæ.

Tae Wei Yuen, space between stars in Leo and Virgo.

Tsze Wei Yuen, circle of perpetual apparition.

Luy Peih Chin, small stars in Aquarius and Pisces.

Le Kung, three groups of two stars each in S. D. Shih (Pegasus).

Tsaou Foo, δ, ε, ζ Cephei.

Neu She, ψ Draconis.

Shang Shoo, A 3687 Draconis.

321   A.D. 1506. *July 31.*

In the reign of the Emperor Woo Tsung, the 1st year of the epoch Ching Tih, 7th moon, day Ke Chow, a star was seen to the west, without the boundary of Tsze Wei. It resembled a great ball. Its colour was a bluish white. After some days it had a small tail. It was seen between S. D. Tsan and Tsing. It gradually lengthened, and appeared like a broom, extending in a north-westerly direction towards Wan Chang.

Emperor Woo Tsung and epoch Ching Tih, 1506-1521: 1st year, 1506: 7th moon, day Ke Chow, July 31.

Tsze Wei, circle of perpetual apparition.

S. D. Tsan determined by a, β and others in Orion.

S. D. Tsing determined by γ, ε, λ, μ and others in Gemini.

Wan Chang, θ, υ, ϕ Ursæ Majoris.

322   A.D. 1506. *August* 10.

On the day Kang Tsze a comet was seen. It was bright, and went to the south-east. It was 3 cubits in length. After 3 days it lengthened to 5 cubits. It swept the upper star of Hea Tae, and entered Tae Wei Yuen.

> 1506: 7th moon, day Kang Tsze, August 10.
> Tae Wei Yuen, space between stars in Leo and Virgo.
> Hea Tae, ν, ξ Ursæ Majoris.
> Possibly the same as the preceding.

323   A.D. 1520. *February.*

In the 15th year of the same epoch, 1st moon, a comet was seen.

> Ching Tih, 15th year, 1520: 1st moon, February.

324   A.D. 1523. *July.*

In the reign of She Tsung, 2nd year of the epoch Kea Tsing, the 6th moon, there was a comet in Teen She.

> Emperor She Tsung and epoch Kea Tsing, 1522–1566: 2nd year, 1523: 6th moon, July.
> Teen She, space within Serpens.

325   A.D. 1531. *August* 5.

In the 10th year of the same epoch, the intercalary 6th moon, day Yih Sze, a comet was seen in the east of S. D. Tsing. Its length was about 1 cubit. It swept the first star in Heen Yuen. The tail gradually increased in length. It went on to S. D. Yih. It was then about 7 cubits in length. It swept Teen Tsan to the north-east. It entered Tae Wei Yuen and swept Lang Wei. It passed through the degrees of S. D. Keo, going to the south-east. It swept the second star to the north of S. D. Kang. It gradually lessened, and after 34 days it disappeared.

> Kea Tsing, 10th year, 1531: intercalary 6th moon, day Yih Sze, August 5.
> S. D. Tsing determined by γ, ε, λ, &c. Geminorum.
>   Yih determined by α and others in Crater.
>   Kang determined by ι, κ, χ, θ Virginis.
> Heen Yuen, α and other stars in Leo and Leo Minor.
> Teen Tsan, ψ Ursæ Majoris.   Lang Wei, Coma Berenices.
> Tae Wei Yuen, space between stars in Leo and Virgo.

326   A.D. 1532. *September* 2.

In the 11th year of the same epoch, the 8th moon, day Ke Maou, a comet was seen in the east of S. D. Tsing. It was about a cubit in length. It afterwards went to the north-east. It passed through Teen Tsin. It gradually increased to about 10 cubits in length. It swept the star Choo in Tae Wei Yuen, and Teen Mun in S. D. Keo. In the 12th moon, day Kea Seuh, after having been visible for 115 days, it disappeared.

Kea Tsing, 11th year, 1532: 8th moon, day Ke Maou, September 2; day Kea Seuh, December 26.

S. D. Tsing determined by γ, ε, λ, μ, &c. Geminorum.

S. D. Keo determined by Spica and ξ Virginis.

Tae Wei Yuen, space between stars in Leo and Virgo.

Teen Tsin, α and other stars in Cygnus.

Teen Mun, stars between Spica and γ Hydræ.

327   A.D. 1533. *July* 1.

In the 12th year of the same epoch, 6th moon, day Sin Sze, a comet was seen in Woo Chay. Its length was 5 cubits. It swept Tse Ling and Teen Ta Tseang Keuen. It gradually increased to about 10 cubits. It swept Ko Taou and passed over Tang Shay. In the 8th moon, day Woo Seuh, it disappeared.

Kea Tsing, 12th year, 1533: 6th moon, day Jin Sze, July 1; 8th moon, day Woo Seuh, September 16.

Woo Chay, α, β, θ, κ Aurigæ and β Tauri.

Teen Ta Tseang Keuen, γ Andromedæ, Triangulum, and stars near.

Ko Taou, ν, ξ, o, π Cassiopeiæ.

Tang Shay, ω Cygni and stars near in Lacerta, &c.

328   A.D. 1539. *April* 30.

In the 18th year of the same epoch, 4th moon, day Kang Seuh, a comet was seen. It was about 3 cubits in length. It was bright, and pointed towards the south-west. It swept the 8th star of Heen Yuen. After 10 days it disappeared.

Kea Tsing, 18th year, 1539: 4th moon, day Kang Seuh, April 30.

Heen Yuen, α and other stars in Leo and Leo Minor.

329   A.D. 1554. *June* 23.

In the 33rd year of the same epoch, the 5th moon, day Kwei Hae, a comet was seen near to Teen Keuen. It entered Wan Chang. It came near the star Shuh. It was visible for 27 days, and then disappeared.

Kea Tsing, 33rd year, 1554: 5th moon, day Kwei Hae, June 23.

Teen Keuen, δ Ursæ Majoris.

Wan Chang, θ, ν, φ Ursæ Majoris.

Shuh, α Serpentis.

330   A.D. 1556. *March* 1.

In the 35th year of the same epoch, the 1st moon, day Kang Shin, a comet was seen near Tsin Heen. It was more than a cubit in length. It pointed towards the

south-west, and gradually increased in length to about 3 cubits. It swept Tae Wei
Yuen, to the north-east of Tsze Seang. It entered Tsze Wei Yuen and came near to
Teen Chwang. On the 2nd day of the 4th moon it disappeared.

Kea Tsing, 35th year, 1556: 1st moon, day Kang Shin, March 1. The 2nd
day of the 4th moon, about May 27.

Tsze Wei Yuen, circle of perpetual apparition.
Tae Wei Yuen, space between Leo and Virgo.
Tsin Heen, ψ, χ and others in Virgo.
Tsze Seang, δ Virginis.
Teen Chwang, stars near θ Draconis.

331   A.D. 1557. *October* 10.

In the 36th year of the same epoch, 9th moon, day Woo Shin, a comet was seen in
Teen She Yuen, near Le Sze, to the north-west. It remained until the 23rd year of the
10th moon, when it disappeared.

Kea Tsing, 36th year, 1557: 9th moon, day Woo Shin, October 10.
Teen She Yuen, space bounded by Serpens.
Le Sze, λ Ophiuchi and small stars near.

332   A.D. 1569. *November* 9.

In the reign of Muh Tsung, the 3rd year of the epoch Lung King, the 10th moon,
day Sin Chow, the 1st day of the moon, a comet was seen in Teen She Yuen. It pointed
to the north-east. On the day Kang Shin it disappeared.

Emperor Muh Tsung and epoch Lung King, 1567–1572: 3rd year, 1569:
10th moon, day Sin Chow, November 9; Kang Shin, November 28.
Teen She Yuen, space bounded by Serpens.

333   A.D. 1577. *November* 14.

In the reign of Shin Tsung, 5th year of the epoch Wan Le, 10th moon, day Woo
Tsze, a comet was seen to the south-west. Its colour was a bluish white. Its length
was estimated at 10 cubits. Its vapour (tail) was perfectly white. From the S. D.
Wei and Ke it passed over S. D. Tow and New. It approached near to S. D. Neu.
It was visible for 1 moon, and then disappeared.

Emperor Shin Tsung and epoch Wan Le, 1573–1619: 5th year, 1577: 10th
moon, day Woo Tsze, November 14.
S. D. Wei determined by γ, δ, ε in Scorpio.
Ke determined by γ, δ, ε, &c. Sagittarii.
Tow determined by ζ, τ, σ, &c. Sagittarii.
New determined by a, β, &c. Capricorni.
Neu determined by ε, μ, &c. Aquarii.

334   A.D. 1580. *October* 1.

In the 8th year of the same epoch, 8th moon, day Kang Shin, a comet was seen in the south-east. It increased in size a little every night. It passed along Ho Han. It was seen altogether for 70 days, and then disappeared.

> Epoch Wan Le, 8th year, A.D. 1580 : 8th moon, day Kang Shin, October 1.
> Ho Han, the Milky Way.

335   A.D. 1582. *May* 20.

In the 10th year of the same epoch, the 4th moon, day Ping Shin, a comet was seen in the north-west. It resembled a folded piece of dyed silk. The tail pointed to Woo Chay. It was visible for about 20 days, and then disappeared.

> Epoch Wan Le, 10th year, 1582 : 4th moon, day Ping Shin, May 20.
> Woo Chay, a, β, θ, ι Aurigæ and β Tauri.

336   A.D. 1585. *October* 3.

In the 13th year of the same epoch, 9th moon, day Woo Tsze, a comet appeared near Yu Lin. It was more than a cubit in length. Each night it went to the east. It gradually lessened, and in the 10th moon, day Kwei Yin, it disappeared.

> Wan Le, 13th year, 1585 : 9th moon, day Woo Tsze, October 3 ; day Kwei Yew, November 17.
> Yu Lin, δ, τ, a Aquarii.

337   A.D. 1591. *April* 3.

In the 19th year of the same epoch, the 3rd moon, day Ping Shin, there was a star like a broom in the north-west, about a cubit in length. It passed over S. D. Wei, Shih, and Peih. Its length was then 2 cubits. In the intercalary 3rd moon, on the day Ping Yin, the 1st day of the moon, it entered S. D. Lew.

> Wan Le, 19th year, 1591 : 3rd moon, day Ping Shin, April 3 ; intercalary 3rd moon, day Ping Yin, April 13.
> S. D. Wei determined by a Aquarii, &c.
>> Shih determined by a, β Pegasi and others near.
>> Peih determined by γ Pegasi and a Andromedæ.
>> Lew determined by a, β, γ Arietis.

338   A.D. 1593. *July* 20.

In the 21st year of the same epoch, 7th moon, day Yih Maou, a comet was seen in the eastern part of S. D. Tsing. On the day Yih Hae it went the contrary way, entered Tsze Wei Yuen, and approached closely to Hwa Kae.

z

Wan Le, 21st year, 1593: 7th moon, day Yih Maou, July 20; day Yih Hae, August 9.

S. D. Tsing determined by γ, ε, λ, μ, &c. Geminorum.

Tsze Wei Yuen, circle of perpetual apparition.

Hwa Kae, small stars in Cassiopeia and Camelopardalis (uncertain).

339 A.D. 1596. *July* 26.

In the 24th year of the same epoch, 7th moon, day Ting Chow, a comet was seen in the north-west. It resembled a round ball. It entered S. D. Yih. It was about a cubit in length. Its course was towards the north-west.

Wan Le, 24th year, 1596: 7th moon, day Ting Chow, July 26.

S. D. Yih determined by α and other stars in Crater.

340 A.D. 1607. *September* 11.

In the 35th year of the same epoch, the 8th moon, day Sin Yew, the 1st day of the moon, a comet was seen in the eastern part of S. D. Tsing. It pointed to the south-west. It went slowly to the north-west. On the day Jin Woo it passed from S. D. Fang into S. D. Sin and disappeared.

Wan Le, 35th year, 1607: 8th moon, day Sin Yew, September 11; day Jin Woo, October 2.

S. D. Tsing determined by γ, ε and other stars in Gemini.

Fang determined by β, δ and others in Scorpio.

Sin determined by Antares and others in Scorpio.

341 A.D. 1618. *November* 16.

In the 46th year of the same epoch, 10th moon, day Yih Chow, a comet appeared in S. D. Te. Its length was about 10 cubits. It pointed to the south-east. It gradually pointed to the north-west. It swept over the star Tae Yang Shoo. It entered S. D. Kang, about a degree to the north-west. It swept Pih Tow, the stars Seuen and Ke, Wan Chang, and Woo Chay. It passed off Tsze Wei Yuen. In the 11th moon, day Kea Shin, it disappeared.

Wan Le, 46th year, 1618: 10th moon, day Yih Chow, November 16; day Kea Shin, December 25.

S. D. Te determined by α, β, γ, &c. Libræ.

S. D. Kang determined by ι, κ, λ, θ Virginis.

Tsze Wei Yuen, circle of perpetual apparition.

Tae Yang Shoo, χ Ursæ Majoris.

Pih Tow, the seven bright stars in Ursa Major.

Seuen, β Ursæ Majoris.   Ke, γ Ursæ Majoris.

Wan Chang, θ, υ, φ Ursæ Majoris.

Woo Chay, α, β, θ, &c. Aurigæ, and β Tauri.

342   A.D. 1619. *February.*

In the 47th year of the same epoch, 1st moon, a comet was seen in the south-east. Its length was estimated at 100 cubits. Its luminous envelope pointed downwards: the end was curved and pointed.

    Wan Le, 47th year, 1619: 1st moon, February.

343   A.D. 1639.

In the reign of Chwang Le, 12th year of the epoch Tsung Ching, a comet was seen in the degrees of S. D. Tsan.

    Emperor Chwang Le and epoch Tsung Ching, 1628–1644: 12th year, 1639.
    S. D. Tsan determined by a, β, γ, δ, &c. Orionis.

344   A.D. 1640. *December* 12.

In the 13th year of the same epoch, 10th moon, day Ping Seuh, a comet was seen.

    Tsung Ching, 13th year, 1640: 10th moon, day Ping Seuh, December 12.

THE Observations that follow form a separate section in the 'She Ke,' in which they are termed those of Temporary or Strange Stars. Some of these are undoubtedly meteors, and have consequently been omitted here, where there was any reason to believe them comets, or where there was anything particularly interesting relating to them they have been retained. They are all of the Ming dynasty.

345   A.D. 1376. *June* 22.

In the reign of Tae Tsoo, 9th year of the epoch Hung Woo, the 6th moon, day Woo Tsze, there was a great star resembling a round ball. Its colour was white. It was situated in Teen Tsang. It crossed Wae Ping and Keuen She. It entered Tae Wei Yuen. It swept Wan Chang and pointed towards Nuy Shoo. It entered into S. D. Chang. In the 7th moon, day Yih Hae, it disappeared.

    Emperor Tae Tsoo and epoch Hung Woo, 1368–1398: 9th year, 1376: 6th moon, day Woo Tsze, June 22; 7th moon, day Yih Hae, August 8.
    S. D. Chang determined by κ, λ, μ, &c. Hydræ.
    Teen Tsang, ι, θ, η, in Cœtus.   Keuen She, ν Persei.
    Tsze Wei Yuen, circle of perpetual apparition.
    Wan Chang, θ, ϕ, ν Ursæ Majoris.
    Teen Shoo, or Nuy Shoo, δ and other small stars in Draco.

346   A.D. 1378. *September* 26.

In the 11th year of the same epoch, 9th moon, day Kea Seuh, a star was seen to the north-east, in Woo Chay. It put forth a tail about 10 cubits in length. It passed over Nuy Keae. It entered Tsze Wei Kung. It swept the five stars of Pih Keih. It passed over Shaou Tsae of Tung Yuen. It entered Teen She Yuen, and remained there until the 10th moon, day Ke Wei; when, on account of cloudy weather, it could no longer be seen.

    Hung Woo, 11th year, 1378 : 9th moon, day Kea Seuh, September 26 ; Ke Wei, November 10.

    Woo Chay, *a*, *β*, &c. Aurigæ, and *β* Tauri.

    Nuy Keae, *τ* and others in Ursa Major.

    Tsze Wei Kung, circle of perpetual apparition.

    Pih Keih, Polaris, and others near.

    Shaou Tsae, *η* Draconis.

    Teen She Yuen, space bounded by Serpens.

347   A.D. 1385. *October 23.*

    In the 18th year of the same epoch, 9th moon, day Yin Yew, a comet was seen in Tae Wei Yuen. It came very near to Yew Chih Fa, and passed out by Twan Mun. On the day Yih Yew it entered S. D. Yih. Its length was then about 10 cubits. In the 10th moon, day Kang Yin, it entered Keen Mun, and swept Teen Meaou.

    Hung Woo, 18th year, 1385 : 9th moon, day Woo Yin, October 23 ; Yih Yew, October 30 : 10th moon, day Kang Yin, November 4.

    S. D. Yih determined by *a*, *β* and others in Crater.

    Tae Wei Yuen, space between stars in Leo and Virgo.

    Yew Chih Fa, *β* Virginis.

    Twan Mun, space between *β* and *η* Virginis.

    Keen Mun, stars in Hydra, between Crater and Corvus.

    Teen Meaou, probably stars in Argo Navis.

348   A.D. 1388. *March 29.*

    In the 21st year of the same epoch, 2nd moon, day Ping Seuh, a star appeared in the eastern part of S. D. Peih.

    Hung Woo, 1388 : 2nd moon, day Ping Yin, March 29.

    S. D. Peih determined by *γ* Pegasi and *a* Andromedæ.

349   A.D. 1430. *September 9.*

    In the reign of Seuen Tsung, the 5th year of epoch Seuen Tih, the 8th moon, day Kang Yin, a star was seen near Nan Ho. It resembled a large round ball. Its colour was a dark blue. It was seen altogether for 26 days, and then disappeared.

    Emperor Seuen Tsung and epoch Seuen Tih, 1426–1435 : 5th year, 1430 : 8th moon, day Kang Yin, September 9.

    Nan Ho, *a*, *β*, &c. Canis Minoris.

350   A.D. 1430. *November* 14.

In the 10th moon of the same year, day Ping Shin, an extraordinary star was seen to the south of Wae Ping. Its course was to the south-east. It crossed Teen Tsang and Teen Yu. It was visible for 8 days, and then disappeared.

> Seuen Tih, 5th year, 1430: 10th moon, day Ping Shin, November 14.
> Wae Ping, δ, ε, μ, ν Piscium.
> Teen Tsang, ν, θ, η in Cœtus.
> Teen Yu, small stars below Cœtus in Fornax.

351   A.D. 1431. *January* 3.

In the 12th moon of the same year, day Ting Hae, a star like a round ball was seen near Kew Yew. Its colour was a yellowish white. It was not bright. After 15 days it disappeared.

> 1430: 12th moon, day Ting Hae: 1431, January 3.
> Kew Yew, μ, ω, &c. Eridani.

352   A.D. 1453. *January* 4.

In the reign of King Te, the 3rd year of the epoch King Tae, the 11th moon, day Kwei Wei, there was a star seen in S. D. Kwei, near Tseih She Ke. It went very slowly to the west.

> King Te appears to have been a regent during the captivity of the Emperor Ying Tsung. His rule and epoch King Tae, 1450–1454: 3rd year, 1452: 11th moon, day Kwei Wei, 1453, January 3.
> S. D. Kwei determined by γ, δ, η, θ Cancri.
> Tseih She Ke, Præsepe in Cancer.

353   A.D. 1458. *December* 24.

In the reign of Ying Tsung, 2nd year of the epoch Teen Shun, 11th moon, day Kwei Maou, there was a star seen in S. D. Sing. Its colour was white. It went westward until the day Ping Woo, when its body faded away. Its appearance was like meal, or the refuse of silk. Its place was near Heen Yuen. On the day Kang Seuh it produced a tail $\frac{5}{10}$ths of a cubit in length. It invaded the north-west star of Kwan Wei. In the 12th moon, day Jin Seuh, it disappeared in the eastern part of S. D. Tsing.

> Emperor Ying Tsung and epoch Teen Shun, 1457–1464: 2nd year, 1458: 11th moon, day Kwei Maou, December 24; day Ping Woo, December 27; day Kang Seuh, December 31: 12th moon, day Jin Seuh, January 12, 1459.
> S. D. Sing determined by σ, τ, &c. Hydræ.
> S. D. Tsing determined by γ, ε, λ, μ, &c. Geminorum.
> Heen Yuen, a, γ and other stars in Leo and Leo Minor.
> Kwan Wei, λ, μ and other stars in Cancer.

A A

354   A.D. 1461. *June 29.*

In the 5th year of the same epoch, 6th moon, day Jin Shin, a star resembling white meal was seen near Tsung Ching, in Teen She Yuen. On the day Yih Wei it changed into a white vapour and disappeared.

> Teen Shun, 5th year, 1461 : 6th moon, day Jin Shin, June 29 ; day Yih Wei, August 2.
> Teen She Yuen, space bounded by Serpens.
> Tsung Ching, β, γ Ophiuchi.

355   A.D. 1462. *June 29.*

In the 6th year of the same epoch, 6th moon, day Ping Yin, a star was seen near the star Tsih. Its colour was a bluish white. It entered Tsze Wei Yuen. It invaded Teen Laou. On the day Kwei Wei it was beneath Chung Tae. Its form gradually faded away.

> Teen Shun, 6th year, 1462 : 6th moon, day Ping Yin, June 29 ; Kwei Wei, July 16.
> Tsze Wei Yuen, circle of perpetual apparition.
> Tsih, δ Cassiopeiæ.
> Teen Laou, ω and others in Ursa Major.
> Chung Tae, λ, μ Ursæ Majoris.

356   A.D. 1491. *January 19.*

In the reign of Heaou Tsung, 3rd year of epoch Hung Che, 12th moon, day Ting Sze, a star was seen in Teen She Yuen. It went to the south-east. On the day Woo Shin it was seen beneath Teen Tsang. It gradually went towards S. D. Peih.

> Emperor Heaou Tsung and epoch Hung Che, 1488–1505 : 3rd year, 1490 : 12th moon, day Ting Sze, January 19, 1491 ; Woo Shin, January 30.
> Teen She Yuen, space bounded by Serpens.
> Teen Tsang, ι, η, θ, &c. Cœti.
> S. D. Peih determined by γ Pegasi and α Andromedæ.

357   A.D. 1495. *January 7.*

In the 7th year of the same epoch, 12th moon, day Ping Yin, a star was seen near Teen Keang. It went slowly towards S. D. Tow until the 8th year, 1st moon, day Kang Seuh, when it entered S. D. Wei.

> Hung Che, 7th year, 1494 : 12th moon, day Ping Yin, January 7, 1495 : 8th year, 1st moon, day Kang Seuh, 1495, February 20.
> S. D. Tow determined by ζ, τ, σ, &c. Sagittarii.
> S. D. Wei determined by α Aquarii and θ, ε Pegasi.
> Teen Keang, θ and others in Ophiuchus.

358 A.D. 1499. *August* 16.

In the 12th year of the same epoch, 7th moon, day Woo Shin, a star was seen near the star Tsung in Teen She Yuen. It entered the eastern boundary of Tsze Wei Yuen. It passed Shaou Tsae and Shang Shoo. It touched Tae Tsze and How Kung. It passed out of the western boundary near Shaou Foo. It was visible until the 8th moon, day Ke Chow, when it disappeared.

Hung Che, 12th year, 1499: 7th moon, day Woo Shin, August 16; 8th moon, day Ke Chow, September 6.
Teen She Yuen, space bounded by Serpens.
Tsze Wei Yuen, circle of perpetual apparition.
Ta Tsze, γ Ursæ Minoris.
How Kung, β Ursæ Minoris.
Shaou Foo, λ Draconis.

359 A.D. 1502. *November* 28.

In the 15th year of the same epoch, 10th moon, day Woo Shin, a star was seen near Teen Maou, in S. D. Chang. It arrived at S. D. Yih, and having returned again to Chang, on the day Woo Yin it disappeared.

Hung Che, 15th year, 1502: 10th moon, day Woo Shin, November 28 ; day Woo Yin, December 8.
S. D. Chang determined by κ, λ, μ, &c. Hydræ.
S. D. Yih determined by a and others in Crater.
Teen Maou, stars in Argo Navis.

360 A.D. 1521. *February* 7.

In the reign of Woo Tsung, the 16th year of the epoch Ching Tih, the 1st moon, day Kea Yin, the 1st day of the moon, there was a star in the south-east. It resembled a changing flame of fire, of a white colour, and was from 6 to 7 cubits in length. It crossed the heavens from east to west, and was dissipated.

Emperor Woo Tsung and epoch Ching Tih, 1506-1521: 16th year, 1521: 1st moon, day Kea Yin, February 7.

361 A.D. 1529. *February* 5.

In the reign of She Tsung, 8th year of the epoch Kea Tsing, the 1st moon, on the day of Leih Chun, a long star extended across the heavens. The same occurred in the 7th moon.

Emperor She Tsung and Kea Tsing, 1522-1566: 8th year, 1529: 1st moon, day of Lei Chun. Leih Chun is the 3rd of the 24 divisions of the year, being that of the beginning of spring: it answers to our February 5. 7th moon, August.

362   A.D. 1532. *March* 9.

In the 11th year of the same epoch, the 2nd moon, day Jin Woo, a star was seen in the south-east. Its colour was a bluish white. It had a tail. After 19 days it disappeared.

> Kea Tsing, 11th year, 1532 : 2nd moon, day Jin Woo, March 9.

363   A.D. 1534. *June* 12.

In the 13th year of the same epoch, 5th moon, day Ting Maou, the 1st day of the moon, a star was seen in Tang Shay. It passed through Teen Ke and entered Ko Taou. On the 24th day it disappeared.

> Kea Tsing, 13th year, 1534 : 5th moon, day Ting Maou, June 12.
> Tang Shay, stars in Cygnus, Lacerta, and Andromeda.
> Teen Ke, θ, ρ, σ and others in Andromeda.
> Ko Taou, ν, ξ, o and others in Cassiopeia.

364   A.D. 1536. *March* 24.

In the 15th year of the same epoch, the 3rd moon, day Woo Woo, a star was seen near Teen Kae. It went to the east. It passed through Teen Choo to the west. It entered Teen Han, and in the 4th moon, day Jin Shin, it disappeared.

> Kea Tsing, 15th year, 1536 : 3rd moon, day Woo Woo, March 24 ; 4th moon, day Jin Shin, April 27.
> Teen Kae, β, γ and others in Draco.
> Teen Choo, δ and others in Draco.
> Teen Han, the Milky Way.

365   A.D. 1545. *December* 26.

In the 24th year of the same epoch, the 11th moon, day Jin Woo, a star appeared in Teen Kae. It entered S. D. Ke. It turned and went to the north-east. At the end of the moon it disappeared.

> Kea Tsing, 24th year, 1545 : 11th moon, day Jin Woo, December 26.
> S. D. Ke determined by γ, δ, ε Sagittarii.
> Teen Kae, β, γ, &c. in Draco.

366   A.D. 1578. *February* 22.

In the reign of Shin Tsung, 6th year of epoch Wan Le, 1st moon, day Woo Shin, a great star resembling the Sun appeared in the west, surrounded by a number of stars, all in the west.

> Emperor Shin Tsung and epoch Wan Le, 1573–1617 : 6th year, 1578 : 1st moon, day Woo Shin, February 22.

367   A.D. 1584.   *July* 1.

In the 12th year of the same epoch, 6th moon, day Ke Yew, a star appeared in
S. D. Fang.

> Wan Le, 12th year: 6th moon, day Ke Yew, July 1, 1584.
> S. D. Fang determined by β, δ, π, ρ in Scorpio.

368   A.D. 1604.   *September* 30.

In the 32nd year of the same epoch, the 9th moon, day Yih Chow, a star was seen
in the degrees of S. D. Wei.   It resembled a round ball.   Its colour was a reddish
yellow.   It was seen in the south-west until the 10th moon, when it was no longer
visible.   In the 12th moon, day Sin Yew, it again appeared in the south-east, in S. D.
Wei.   The next year, in the 2nd moon, it gradually faded away.   In the 8th moon, day
Ting Maou, it disappeared.

> Wan Le, 32nd year, 1604: 9th moon, day Yih Chow, September 30; 10th
> moon, November; 12th moon, day Sin Yew, 1605, January 14: 33rd year, 1605:
> 2nd moon, day Ting Maou, March 21.
> S. D. Wei determined by ε, μ, ν and others in Scorpio.
>
> Biot has S. D. Fang instead of the second S. D. Wei.   S. D. Fang is deter-
> mined by β, δ, π and others in Scorpio.   It is, however, Wei in the 'She Ke.'

369   A.D. 1609.

In the 37th year of the same epoch a great star was seen in the south-west.   The
tail had four rays.

> Wan Le, 37th year, 1609.

370   A.D. 1618.   *November* 24.

In the 46th year of the same epoch, the 9th moon, day Yih Maou, a white vapour
was seen in the south-east.   It was about a cubit in width and 20 cubits in length.   It
extended from the east to the west of S. D. Chin.   It entered S. D. Yih, and after 19
days it disappeared.

> S. D. Chih determined by β, &c. Corvi.
> Yih, a and others in Crater.

371   A.D. 1618.   *December* 5.

In the 11th moon of the same year, day Ping Yin, in the morning, a star like a
white flower was seen.

> 1618: 11th moon, day Ping Yin, December 5.

B B

372   A.D. 1621.   *May* 12.

In the reign of He Tsung, the 1st year of the epoch Teen Ke, the 4th moon, day Kwei Yew, a reddish star was seen in the east.

Emperor He Tsung and epoch Teen Ke, 1621–1627: 1st year, 1621: 4th moon, day Kwei Yew, May 12.

APPENDIX,

CONSISTING OF

TABLES

FOR

REDUCING CHINESE TIME TO EUROPEAN RECKONING,

AND

A CHINESE CELESTIAL ATLAS.

Chinese Chronological Tables;

SHOWING

THE SUCCESSION OF THE DYNASTIES AND EMPERORS,

FROM THE EARLIEST PERIOD TO THE PRESENT TIME.

.*. These Tables are required for finding the Year of any occurrence. The method of using these and the subsequent Tables is fully explained in the Introductory Remarks.

SUCCESSION OF THE DYNASTIES,

FROM THE ACCESSION OF THE HEA TO THAT OF THE PRESENT DYNASTY,
THE TSING.

Dynasties.		Date.		Dynasties.		Date.
		B. C.				A. D.
夏	Hea	2205		陳	Chin	557
商	Shang	1766		隋	Suy	589
周	Chow	1122		唐	Tang	618
東周	Tung Chow	696		後梁	How Leang	907
泰	Tsin	255		後唐	How Tang	923
漢	Han	206		後晉	How Tsin	936
東漢	Tung Han	A. D. 25		後漢	How Han	947
屬漢	Shuh Han	221		後周	How Chow	951
晉	Tsin	265		宋	Sung	960
東晉	Tung Tsin	317		元	Yuen	1280
宋	Sung	420		明	Ming	1368
齊	Tse	479		清	Tsing	1644
梁	Leang	502				

Chinese Chronology may be arranged under Three Divisions—the Fabulous Period, the Uncertain Period, and that which they consider as certain.

THE FABULOUS PERIOD.

Emperor's Name.		Reigned Years.
盤 古	Pwan Koo	The First Man.
天 皇 氏	Teen Hwang She	18,000
地 皇 氏	Te Hwang She	18,000
人 皇 氏	Jin Hwang She	45,000

THE UNCERTAIN PERIOD.

三 皇 SAN HWANG. THE THREE HWANGS.

Emperor's Name.		Date.	Reigned Years.
		B. C.	
伏 羲	Fuh He	3328	115
神 農	Shin Nung	3213	140
帝 臨	Te Lin	3073	80
帝 承	Te Ching	2993	60
帝 明	Te Ming	2933	49
帝 宜	Te E	2884	45
帝 來	Te Lae	2839	48
帝 裏	Te Le	2791	43
帝 榆	Te Yu	2748	50
皇 帝	Hwang Te	2698	101

The three Hwangs are Fuh He, Shin Nung, and Hwang Te.

From the 1st year of the 1st epoch, 2637 B.C., being the 60th year of Hwang Te, the Chronology is considered as certain.

五 帝 Woo Te. The Five Te's.

(THE WORDS HWANG AND TE ARE IMPERIAL TITLES.)

Emperor's Name.		Date.	Reigned Years.
少 昊	Shaou Haou	B. C. 2597	84
顓 頊	Chuen Kuh	2513	78
帝 嚳	Te Kwuh	2435	79
帝 堯	Te Yaou	2356	101
帝 舜	Te Shun	2255	50

夏 朝 Hea Chaou. The Hea Dynasty, B.C. 2205–1765.

		Date	Reigned Years
大 禹	Ta Yu	2205	8
帝 啟	Te Ke	2197	9
太 康	Tae Kung	2188	29
仲 康	Chung Kang	2159	13
王 相	Wang Seang	2146	28
少 康	Shaou Kang	2118	61
王 杼	Wang Choo	2057	17
王 槐	Wang Hwae	2040	26
王 芒	Wang Mang	2014	18
王 泄	Wang See	1996	16
王 不 降	Wang Puh Keang	1980	59
王 扃	Wang Shang	1921	21
王 廑	Wang Kin	1900	21
王 孔 甲	Wang Kung Kea	1879	31
王 皇	Wang Kaou	1848	11
王 發	Wang Fa	1837	19
桀 癸	Kee Kwei	1818	53

商 朝 SHANG CHAOU.

THE SHANG DYNASTY, B.C. 1766–1122.

Emperor's Name.		Date.	Reigned Years.
成湯	Ching Tang	B.C. 1766	13
太甲	Tae Kea	1753	33
沃丁	Yuh Ting	1720	29
太庚	Tae Kang	1691	25
小甲	Seaou Kea	1666	17
雍已	Yung Ke	1649	12
太戊	Tae Woo	1637	75
仲丁	Chung Ting	1562	13
外壬	Wae Jin	1549	15
河亶甲	Ho Tan Kea	1534	9
祖乙	Tsoo Yih	1525	19
祖辛	Tsoo Sin	1506	17
沃甲	Yuh Kea	1490	29
祖丁	Tsoo Ting	1465	32
南庚	Nan Kang	1433	} 25
陽甲	Yang Kea	1408	
盤庚	Pwan Kang	1401	28
小辛	Seaou Sin	1373	21
小乙	Seaou Yih	1352	28
武丁	Woo Ting	1324	59
祖庚	Tsoo Kang	1265	7
祖甲	Tsoo Kea	1258	33
廩辛	Lin Sin	1225	6
庚丁	Kang Ting	1219	21

Emperor's Name.		Date.	Reigned Years.
武 乙	Woo Yih	B.C. 1198	4
太 丁	Tae Ting	1194	3
帝 乙	Te Yih	1191	37
紂 辛	Chow Sin	1154	32

周 朝 CHOW CHAOU.

THE CHOW DYNASTY, B.C. 1122–254. 868 YEARS.

武 王	Woo Wang	1122	7
成 王	Ching Wang	1115	37
康 王	Kang Wang	1078	26
昭 王	Chaou Wang	1052	51
穆 王	Mo Wang	1001	55
共 王	Kung Wang	946	12
懿 王	E Wang	934	25
孝 王	Heaou Wang	909	15
夷 王	E Wang	894	16
厲 王	Le Wang	878	51
宣 王	Seuen Wang	827	46
幽 王	Yew Wang	781	11
平 王	Ping Wang	770	51
桓 王	Hwan Wang	719	23

東 周 TUNG CHOW.

莊 王	Chwang Wang	696	15
釐 王	Le Wang	681	5

Emperor's Name.	Date.	Reigned Years.
惠 王 Hwuy Wang	B.C. 676	25
襄 王 Seang Wang	651	33
頃 王 King Wang	618	6
匡 王 Kwang Wang	612	6
定 王 Ting Wang	606	21
簡 王 Keen Wang	585	14
靈 王 Ling Wang	571	27
景 王 King Wang	544	25
敬 王 King Wang	519	44
元 王 Yuen Wang	475	7
貞 定 王 Ching Ting Wang	468	28
考 王 Kaou Wang	440	15
威 烈 王 Wei Leĕ Wang	425	24
安 王 Gan Wang	401	26
烈 王 Leĕ Wang	375	7
顯 王 Heen Wang	368	48
愼 靚 王 Shin Tsing Wang	320	6
赧 王 Nan Wang	314	59
東 朝 王 Tung Chow Wang	255	7

泰 朝 TSIN CHAOU.   THE TSIN DYNASTY, B.C. 225–205.

昭 襄 王 Shaou Seang Wang	255	5
孝 文 王 Haou Wan Wang	250	10
莊 襄 王 Chwang Seang Wang	240	4
始 皇 帝 Che Hwang Te	236	37
二 世 皇 帝 Urh She Hwang Te	209	3

漢 朝 Han Chaou. Han Dynasty, b.c. 206 *to* a.d. 264.

西 漢 Se Han. Western Han.

Emperor's Name.	Epoch.	Duration of Epoch. (B.C.)	Reigned Years.	Duration of Reign. (B.C.)
高帝 Kaou Te			12	206 to 195
惠帝 Hwuy Te			7	194 188
高后 Kaou How			8	187 180
文帝 Wan Te	None for 16 yrs.	179 to 164		
	後元 How Yuen, 1st epoch }	163 157	23	179 157
景帝 King Te	None for 7 yrs.	156 150		
	中元 Chung Yuen	149 144		
	後元 How Yuen	143 141	16	156 141
武帝 Woo Te	建元 Keen Yuen	140 135		
	元光 Yuen Kwang	134 129		
	元朔 Yuen Sǒ	128 123		
	元狩 Yuen Show	122 117		
	元鼎 Yuen Ting	116 111		
	元封 Yuen Fung	110 105		
	太初 Tae Choo	104 101		
	天漢 Teen Han	100 97		
	太始 Tae Che	96 93		
	征和 Ching Ho	92 89		
	後元 How Yuen	88 87	54	140 87
昭帝 Chaou Te	始元 Che Yuen	86 81		
	元鳳 Yuen Fung	80 75		
	元平 Yuen Ping	74	13	86 to 74
宣帝 Seuen Te	本始 Pun Che	73 to 70		

Emperor's Name	Epoch	Duration of Epoch (B.C.)	Reigned Years	Duration of Reign (B.C.)
元帝 Yuen Te	地節 Te Tseĕ	69 to 66		
	元康 Yuen Kang	65 62		
	神爵 Shin Tseŏ	61 58		
	五鳳 Woo Fung	57 54		
	甘露 Kan Loo	53 50		
	黃龍 Hwang Lung	49	25	73 to 49
	初元 Choo Yuen	48 44		
	永光 Yung Kwang	43 39		
	建昭 Keen Chaou	38 34		
成帝 Ching Te	竟寧 King Ning	33	16	48 33
	建始 Keen Che	32 29		
	河平 Ho Ping	28 25		
	陽朔 Yang So	24 21		
	鴻嘉 Hung Kea	20 17		
	永始 Yung Che	16 13		
	元延 Yuen Yen	12 9		
哀帝 Gae Te	綏和 Hwan Ho	8 7	26	32 7
	建平 Keen Ping	6 3		
	元壽 Yuen Show	2 1	6	6 1
平帝 Ping Te	元始 Yuen Che	A.D. 1 5	5	A.D. 1 5
孺子嬰 Joo Sze Ying	居攝 Keu Che	6 7		
	初始 Choo Che	8	3	6 8
王莽 Wang Mang (Usurper.)	建國 Keen Kwo	9 13		
	天鳳 Teen Fung	14 19		
淮陽 Hwae Yang	地皇 Te Hwang	20 22	14	9 22
		23 to 24	2	23 to 24

東漢 TUNG HANG. EASTERN HAN.

Emperor's Name.		Epoch.		Duration of Epoch.		Reigned Years.	Duration of Reign.	
				A. D.			A. D.	
光 武	Kwang Woo	建 武元	Keen Woo	25	to 55			
		中 元	Chung Yuen	56	57	33	25	to 57
明 帝	Ming Te	永 平	Yung Ping	58	75	18	58	75
章 帝	Chang Te	建 初	Keen Choo	76	83			
		元 和	Yuen Ho	84	86			
		章 和	Chang Ho	87	88	13	76	88
和 帝	Ho Te	永 元	Yung Yuen	89	104			
		元 興	Yuen Hing	105		17	89	105
殤 帝	Shang Te	延 平	Yen Ping	106		1	106	
安 帝	Gan Te	永 初	Yung Choo	107	113			
		元 初	Yuen Choo	114	119			
		永 寧	Yung Ning	120				
		建 光	Keen Kwang	121				
		延 光	Yen Kwang	122	125	19	107	125
順 帝	Shun Te	永 建	Yung Keen	126	131			
		陽 嘉	Yang Kea	132	135			
		永 和	Yung Ho	136	141			
		漢 安	Han Gan	142	143			
		建 康	Keen Kang	144		19	126	to 144
冲 帝	Chung Te	永 嘉	Yung Kea	145		1	145	
質 帝	Chih Te	本 初	Pun Choo	146		1	146	
桓 帝	Hwan Te	建 和	Keen Ho	147	149			
		和 平	Ho Ping	150				
		元 嘉	Yuen Kea	151	152			
		永 興	Yung Hing	153	to 154			

Emperor's Name.		Epoch.		Duration of Epoch.	Reigned Years.	Duration of Reign.
				A. D.		A. D.
靈帝	Ling Te	燾嘉 永	Yung Show	155 to 157		
		延熹 永	Yen He	158 166		
		永康 建	Yung Kang	167	21	147 to 167
		寧 建	Keen Ning	168 171		
		熹平 喜	He Ping	172 177		
		光和	Kwang Ho	178 183		
獻帝	Heen Te	中平	Chung Ping	184 189	22	168 189
		初平	Choo Ping	190 193		
		興平	Hing Ping	194 195		
		建安	Keen Gan	196 220	31	190 220

後漢 HOW HAN. THE LATER HAN.

昭烈帝	Chaou Le Te	章武	Chang Woo	221 222	2	221 242
後帝	How Te	建興	Keen Hing	223 237		
		延熙	Yen He	238 257		
		景耀	King Teih	258 262		
		炎興	Yen Hing	263 264	42	223 264

晋朝 TSIN CHAOU. THE TSIN DYNASTY, A.D. 265–419.

西晉 SE TSIN. WESTERN TSIN.

武帝	Woo Te	泰始	Tae Che	265 274		
		咸寧	Han Ning	275 279		
		太康	Tae Kang	280 289	25	265 to 289
惠帝	Hwuy Te	永熙	Yung He	290		
		元康	Yuen Kang	291 to 299		
		永康	Yung Kang	300		

Emperor's Name.		Epoch.		Duration of Epoch.		Reigned Years.	Duration of Reign.	
				A. D.			A. D.	
		永寧	Yung Ning	301				
		太安	Tae Gan	302	to 303			
		永興	Yung Hing	304	305			
		光熙	Kwang He	306		17	290	to 306
懷帝	Hwae Te	永嘉	Yung Kea	307	312	6	307	312
民愍帝	Min Te	建興	Keen Hing	313	316	4	313	316

東晉 TUNG TSIN. EASTERN TSIN.

元帝	Yuen Te	建武	Keen Woo	317				
		大興	Ta Hing	318	321			
		永昌	Yung Chang	322		6	317	322
明帝	Ming Te	太寧	Tae Ning	323	325	3	323	325
成帝	Ching Te	咸和	Han Ho	326	334			
		咸康	Han Kang	335	342	17	326	342
康帝	Kang Te	建元	Keen Yuen	343	344	2	343	344
穆帝	Muh Te	永和	Yung Ho	345	356			
		升平	Shing Ping	357	361	17	345	361
哀帝	Gae Te	隆和	Lung Ho	362				
		興寧	Hing Ning	363	365	4	362	365
帝弈	Te Yih	太和	Tae Ho	366	370	5	366	370
簡文帝	Keen Wan Te	咸安	Han Gan	371	372	2	371	372
孝武帝	Heaou Woo Te	寧康	Ning Kang	373	375			
		太元	Tae Yuen	376	396	24	373	396
安帝	Gan Te	隆安	Lung Gan	397	401			
		元興	Yuen Hing	402	404			
		義熙	E He	405	to 418	22	397	to 418
恭帝	Kung Te	元熙	Yuen He	419		1	419	

宋 朝 Sung Chaou. The Sung Dynasty, A.D. 420–478.

Emperor's Name.		Epoch.		Duration of Epoch. A.D.		Reigned Years.	Duration of Reign. A.D.	
帝 武	Woo Te	初 永	Yung Choo	420 to 422		3	420 to 422	
帝 少	Shaou Te	平 景	King Ping	423		1	423	
帝 文	Wan Te	嘉 元	Yuen Kea	424	453	30	424	453
帝 武 孝	Heaou Woo Te	建 孝	Heaou Keen	454	456			
		明 大	Ta Ming	457	464	11	454	464
帝 廢	Fei Te	和 景	King Ho	465		½ year	465	
帝 明	Ming Te	始 泰	Tae Che	465	471			
		豫 泰	Tae Yu	472		8	465	472
王 梧 蒼	Tsang Woo Wang	徽 元	Yuen Hwuy	473	476	4	473	476
帝 順	Shun Te	明 昇	Shing Ming	477	478	2	477	478

齊 朝 Tse Chaou. The Tse Dynasty, A.D. 479–501.

Emperor's Name.		Epoch.		Duration of Epoch.		Reigned Years.	Duration of Reign.	
帝 高	Kaou Te	元 建	Keen Yuen	479	482	4	479	482
帝 武	Woo Te	明 永	Yung Ming	483	493	11	483	493
帝 明	Ming Te	武 建	Keen Woo	494	497	5	494	
		泰 永	Yung Tae	498			498	
後 昏 東	Tung Hwan How	元 永	Yung Yuen	499	500	2	499 to 500	
帝 和	Ho Te	興 中	Chung Hing	501		1	501	

梁 朝 Leang Chaou. The Leang Dynasty, A.D. 502–556.

Emperor's Name.		Epoch.		Duration of Epoch.		Reigned Years.	Duration of Reign.	
帝 武	Woo Te	監 天	Teen Keen	502	519			
		通 晉	Tsin Tung	520	526			
		通 大	Ta Tung	526	527			
		通 大 中	Chung Ta Tung	528 to 534				

Emperor's Name.			Epoch.	Duration of Epoch.		Reigned Years.	Duration of Reign.	
		大 同	Ta Tung	A. D. 535 to 545			A. D.	
		中大 同	Chung Ta Tung	546				
		太 清	Tae Tsing	547	549	47	502 to 549	
簡文帝	Keen Wan Te	大 寶	Ta Paou	550	551	2	550	551
元帝	Yuen Te	承 聖	Ching Shing	552	554	3	552	554
敬帝	King Te	紹 泰	Shaou Tae	555				
		太 平	Tae Ping	556		2	555	556

陳 朝 CHIN CHAOU. THE CHIN DYNASTY, A.D. 557–588.

Emperor's Name.			Epoch.	Duration of Epoch.		Reigned Years.	Duration of Reign.	
武帝	Woo Te	永 定	Yung Ting	557	559	3	557	559
文帝	Wan Te	天 嘉	Teen Kea	560	563			
		天 康	Teen Kang	566		7	560	566
伯宗	Pih Tsung	光 大	Kwang Ta	567	568			
宣帝	Seuen Te	大 建	Ta Keen	569	582	14	569	582
後王	How Wang	至 德	Che Tih	583	586			
		禎 明	Ching Ming	587	588	6	583	588

隋 朝 SUY CHAOU. THE SUY DYNASTY, A.D. 589–617.

Emperor's Name.			Epoch.	Duration of Epoch.		Reigned Years.	Duration of Reign.	
文帝	Wan Te	開 皇	Kae Hwang	589	600			
		仁 壽	Jin Show	601	604	24	589	604
煬帝	Yang Te	大 業	Ta Nee	605 to 616		13	605 to 616	
恭帝	Kung Te	義 寧	E Ning	617		1	617	

F F

唐朝 TANG CHAOU.   THE TANG DYNASTY, A.D. 618–906.

Emperor's Name.	Epoch.	Duration of Epoch. A.D.		Reigned Years.	Duration of Reign. A.D.	
祖 高 Kaou Tsoo	德 武 Woo Tih	618	to 626	9	618	to 626
宗 太 Tae Tsung	觀 貞 Ching Kwan	627	649	23	627	649
高 | * Kaou Tsung	徽 永 Yung Hwuy	650	655			
	慶 顯 Heen King	656	660			
	朔 龍 Lung So	661	663			
	德 麟 Lin Tih	664	665			
	封 乾 Keen Fung	666	667			
	章 總 Tsung Chung	668	669			
	亨 咸 Han Hang	670	673			
	元 上 Shang Yuen	674	675			
	鳳 儀 E Fung	676	678			
	露 調 Kae Teih	679				
	隆 永 Yung Lung	680				
	耀 開 Kae Teih	681				
	淳 永 Yung Shun	682				
	道 宏 Hung Taou	683		34	650	683
中 | Chung Tsung	聖 嗣 Sze Shing	684	704			
	龍 神 Shin Lung	705	706			
	龍 景 King Lung	707	709	26	684	709
睿 | Juy Tsung	雲 景 King Yun	710	711			
	極 太 Tae Keih	712		3	710	712
元 | Yuen Tsung	元 開 Kae Yuen	713	741			
	寶 天 Teen Paou	742	755	43	713 to 755	
肅 | Suh Tsung	德 至 Che Tih	756 to 757				

* Where this mark | occurs, it must be considered as representing the preceding Chinese character.

Emperor's Name.		Epoch.		Duration of Epoch. A.D.		Reigned Years.	Duration of Reign. A.D.	
代宗	Tae Tsung	乾元	Kan Yuen	758 to 759				
		上元	Shang Yuen	760	761			
		寶應	Paou Ying	762		7	756 to 762	
		廣德	Kwang Tih	763	764			
		永泰	Yung Tae	765				
		大歷	Ta Leih	766	779	17	763	779
德	Tih Tsung	建中	Keen Chung	780	783			
		興元	Hing Yuen	784				
		貞元	Ching Yuen	785	804	26	780	804
順	Shun Tsung	永貞	Yung Ching	805		1	805	
憲	Heen Tsung	元和	Yuen Ho	806	820	15	806	820
穆	Muh Tsung	長慶	Chang King	821	824	4	821	824
敬	King Tsung	寶歷	Paou Leih	825	826	2	825	826
文	Wan Tsung	太和	Tae Ho	827	835			
		開成	Kae Ching	836	840	14	827	840
武	Woo Tsung	會昌	Hwuy Chang	841	846	6	841	846
宣	Seuen Tsung	大中	Ta Chung	847	859	13	847	859
懿	E Tsung	咸通	Han Tung	860	873	14	860	873
僖	He Tsung	乾符	Kan Foo	874	879			
		廣明	Kwang Ming	880				
		中和	Chung Ho	881	884			
		光啟	Kwang Ke	885	887			
		文德	Wan Tih	888		15	874 to 888	
昭	Chaou Tsung	龍紀	Lung Ke	889				
		大順	Ta Shun	890	891			
		景福	King Fuh	892 to 893				

Emperor's Name.		Epoch.		Duration of Epoch.	Reigned Years.	Duration of Reign.
				A.D.		A.D.
		乾 寧	Kan Ning	894 to 897		
		光 化	Kwang Hwa	898 900		
		天 復	Teen Fuh	901 903		
		天 袖	Teen Yew	904	16	889 to 904
昭宣帝	Chaou Seuen Te	天 袖	Teen Yew	905 906	2	905 906

五代朝 Woo Tae Chaou, or the Five Short Dynasties, a.d. 907–960.

後梁 How Leang. The Later Leang, a.d. 907–922.

太祖	Tae Tsoo	開 平	Kae Ping	907 910		
		乾 化	Kan Hwa	911 912	6	907 912
末帝	Muh Te	乾 化	Kan Hwa	913 914		
		貞 明	Ching Ming	915 920		
		龍 德	Lung Tih	921 922	10	913 922

後唐 How Tang. The Later Tang, a.d. 923–935.

莊宗	Chwang Tsung	同 光	Tung Kwang	923 925	3	923 925
明宗	Ming Tsung	天 成	Teen Ching	926 929		
		長 興	Chang Hing	930 933	8	926 933
閔帝	Min Te	應 順	Ying Shun	934	3 mths.	
廢帝	Fei Te	清 泰	Tsing Tae	934 935	2	934 935

後晋 How Tsin. The Later Tsin, a.d. 936–946.

高祖	Kaou Tsoo	天 福	Teen Fuh	936 942	7	936 942
出帝	Chuh Te	天 福	Teen Fuh	943 944		
		開 運	Kae Yun	945 to 946	4	943 to 946

後 漢 HOW HAN, 947–950.

Emperor's Name.		Epoch.		Duration of Epoch.	Reigned Years.	Duration of Reign.
				A. D.		A. D.
高祖	Kaou Tsoo	天福	Teen Fuh	947	I	947
隱帝	Yen Te	乾祐	Kan Yew	948 to 950	3	948 to 950

後 周 HOW CHOW, 951–960.

Emperor's Name.		Epoch.		Duration of Epoch.	Reigned Years.	Duration of Reign.
太祖	Tae Tsoo	廣順	Kwang Shun	951	953	3	951	953
世宗	She Tsung	顯德	Heen Tih	954	959	6	954	959
恭帝	Kung Te	… …	… …	… …	½	960

宋 朝 SUNG CHAOU.. (SECOND) SUNG DYNASTY, A.D.. 960–1279.

Emperor's Name.		Epoch.		Duration of Epoch.		Reigned Years.	Duration of Reign.	
太祖	Tae Tsoo	建隆	Keen Lung	960	962			
		乾德	Kan Tih	963	967			
		開寶	Kae Paou	968	975	16	960	975
太宗	Tae Tsung	太平興國	Tae Ping Hing Kwo	976	983			
		雍熙	Yung He	984	987			
		端拱	Twan Kung	988	989			
		淳化	Shun Hwa	990	994			
		至道	Che Taou	995	997	22	976	997
眞		Ching Tsung	咸平	Han Ping	998	1003			
		景德	King Tih	1004	1007			
		大中祥符	Ta Chung Tseang Foo	1008	1016			
		天禧	Teen He	1017	1021			
		乾興	Kan Hing	1022		25	998 to 1022	
仁		Jin Tsung	天聖	Teen Shing	1023	1031			
		明道	Ming Taou	1032 to 1033				

Emperor's Name.		Epoch.		Duration of Epoch.	Reigned Years.	Duration of Reign.
				A. D.		A. D.
宗 英		景 祐	King Yew	1034 to 1037		
		寶 元	Paou Yuen	1038 1039		
		康 定	Kang Ting	1040		
		慶 歷	King Leih	1041 1048		
		皇 祐	Hwang Yew	1049 1053		
		至 和	Che Ho	1054 1055		
		嘉 祐	Kea Yew	1056 1063	41	1023 to 1063
英	Ying Tsung	治 平	Che Ping	1064 1067	4	1064 1067
神	Shin Tsung	熙 寧	He Ning	1068 1077		
		元 豐	Yuen Fung	1078 1085	18	1068 1085
哲	Che Tsung	元 祐	Yuen Yew	1086 1093		
		紹 聖	Shaou Shing	1094 1097		
		元 符	Yuen Foo	1098 1100	15	1086 1100
徽	Hwuy Tsung	建中 靖國	Keen Chung Tsing Kwo	1101		
		崇 寧	Tsung Ning	1102 1106		
		大 觀	Ta Kwan	1107 1110		
		政 和	Ching Ho	1111 1117		
		重 和	Chung Ho	1118		
		宜 和	E Ho	1119 1125	25	1101 1125
欽	Kin Tsung	靖 康	Tsing Kang	1126	1	1126
高	Kaou Tsung	建 炎	Keen Yen	1127 1130		
		紹 興	Shaou Hing	1131 1162	36	1127 1162
孝	Heaou Tsung	隆 興	Sung Hing	1163 1164		
		乾 道	Kan Taou	1165 1173		
		淳 熙	Shun He	1174 1189	27	1163 1189
光	Kwang Tsung	紹 熙	Shaou He	1190 to 1194	5	1190 to 1194

Emperor's Name.		Epoch.		Duration of Epoch.	Reigned Years.	Duration of Reign.
				A. D.		A. D.
寧宗	Ning Tsung	元慶	King Yuen	1195 to 1200		
		泰嘉	Kea Tae	1201 1207		
		禧開	Kae He	1205 1207		
		定嘉	Kea Ting	1208 1224	30	1195 to 1224
理	Le Tsung	慶寶	Paou King	1225 1227		
		定紹	Shaou Ting	1228 1233		
		平端	Twan Ping	1234 1236		
		熙嘉	Kea He	1237 1240		
		祐淳	Shun Yew	1241 1252		
		祐寶	Paou Yew	1253 1258		
		慶開	Kae King	1259		
		定景	King Ting	1260 1264	40	1225 1264
度	Too Tsung	淳咸	Han Shun	1265 1274	10	1265 1274
恭	Kung Tsung	祐德	Tih Yew	1275	1	1275
端	Twan Tsung	炎景	King Yen	1276 1277	2	1276 1277
帝昺	Te Ping	典祥	Tseang Hing	1278 1279	2	1278 1279

元朝 YUEN CHAOU. THE YUEN DYNASTY, A.D. 1280–1367.

Emperor's Name.		Epoch.		Duration of Epoch.	Reigned Years.	Duration of Reign.
世祖	She Tsoo	至元	Che Yuen	1280 1294	15	1280 1294
成宗	Ching Tsung	元貞	Yuen Ching	1295 1296		
		元德	Ta Tih	1297 1307	13	1295 1307
武	Woo Tsung	大至	Che Ta	1308 1311	4	1308 1311
仁	Jin Tsung	皇慶	Hwang King	1312 1313		
		延祐	Yen Yew	1314 1320	9	1312 1320
英	Ying Tsung	至治	Che Che	1321 to 1323	3	1321 to 1323

Emperor's Name.		Epoch.		Duration of Epoch.		Reigned Years.	Duration of Reign.	
				A.D.			A.D.	
泰定帝	Tae Ting Te	泰定	Tae Ting	1324 to 1327			1324 to 1328	
		至和	Che Ho	1328		5		
明宗	Ming Tsung	天歷	Teen Leih	1328		½		
文 ＼	Wan Tsung	天歷	Teen Leih	1328	1329			
		至順	Che Shun	1330	1332	5	1328	1332
宁 ＼	Ning Tsung	… …	… …	1332		1 mo.	1332	
順帝	Shun Te	元統	Yuen Tung	1333	1334			
		至元	Che Yuen	1335	1340			
		至正	Che Ching	1341	1367	35	1333	1367

明朝 Ming Chaou. The Ming Dynasty, a.d. 1368–1644.

Emperor's Name.		Epoch.		Duration of Epoch.		Reigned Years.	Duration of Reign.	
太祖帝	Tae Tsoo	洪武	Hung Woo	1368	1398	31	1368	1398
惠帝	Hwuy Te	文建	Keen Wan	1399	1402	4	1399	1402
成祖宗	Ching Tsoo	永樂	Yung Lo	1403	1424	22	1403	1424
仁 ＼	Jin Tsung	洪熙	Hung He	1425		1	1425	
宣 ＼	Seuen Tsung	宣德	Seuen Tih	1426	1435	10	1426	1435
*英帝宗	Ying Tsung	正統	Ching Tung	1436	1449	14	1436	1449
景 ＼	King Te	景泰	King Tae	1450	1456	7	1450	1456
英 ＼	Ying Tsung	天順	Teen Shun	1457	1468	8	1457	1468
憲 ＼	Heen Tsung	成化	Ching Hwa	1465	1487	23	1465	1487
孝 ＼	Heaou Tsung	弘治	Hung Che	1488	1505	18	1488	1505
武 ＼	Woo Tsung	正德	Ching Tih	1506	1521	16	1506	1521
世 ＼	She Tsung	嘉靖	Kea Tsing	1522	1566	45	1522	1566
穆 ＼	Muh Tsung	隆慶	Lung King	1567	1572	6	1567	1572
神 ＼	Shin Tsung	萬歷	Wan Leih	1573 to 1619		47	1573 to 1619	

* Ying Tsung was taken prisoner by the Tartars in 1450, and restored in 1457, when he changed the epoch to Teen Shun.

Emperor's Name.			Epoch.			Duration of Epoch.	Reigned Years.	Duration of Reign.
光宗	宗	Kwang Tsung	泰	昌	Tae Chang	A. D. 1620	I	A. D. 1620
熹宗	宗	He Tsung	天	啓	Teen Kę	1621 to 1627	7	1621 to 1627
莊烈	烈	Chwang Lee	崇	禎	Tsung Ching	1628 1644	17	1628 1644

清朝 TSING CHAOU. THE TSING DYNASTY, A.D. 1644.

Emperor's Name.			Epoch.			Duration of Epoch.	Reigned Years.	Duration of Reign.
世祖	祖	She Tsoo	順	始	Shun Che	1644 1661	18	1644 1661
聖祖	祖	Shin Tsoo	康	熙	Kang He	1662 1722	61	1662 1722
世宗	宗	She Tsung	雍	正	Yung Ching	1723 1735	13	1723 1735
高	—	Kaou Tsung	乾	隆	Keen Lung	1736 1795	60	1736 1795
仁	—	Jin Tsung	嘉	慶	Kea King	1796 1820	25	1796 1820
宣	—	Seuen Tsung	道	光	Taou Kwang	1821 1850	30	1821 1850
—	—	—	咸	豐	Heen Fung	1851 1862	11	1851 1862
—	—	—	同	治	Tung Che	1863		

THE MINOR DYNASTIES.

魏 WEI, A.D. 220-265.

Emperor's Name.			Epoch.			Duration of Epoch.	Reigned Years.	Duration of Reign.
文帝	帝	Wan Te	黄	初	Hwang Choo	220 226	7	220 226
明帝	帝	Ming Te	太	和	Tae Ho	227 232		
			青	龍	Tsing Lung	233 236		
			景	初	King Choo	237 239	13	227 239
廢帝	帝	Fei Te	正	始	Ching Che	240 248		
			嘉	平	Kea Ping	249 253	14	240 253
少帝	帝	Shaou Te	正	元	Ching Yuen	254 255		
			甘	露	Kan Loo	256 to 259	6	254 to 259

Emperor's Name.		Epoch.		Duration of Epoch.	Reigned Years.	Duration of Reign.
末帝	Mo Te	景元	King Yuen	A. D. 260 to 263		A. D.
		咸熙	Han He	264 265	6	260 to 265

吳 WOO, A.D. 221–280.

Emperor's Name.		Epoch.		Duration of Epoch.	Reigned Years.	Duration of Reign.
大帝	Ta Te	黃武	Hwang Woo	221 228		
		黃龍	Hwang Lung	229 231		
		嘉禾	Kea Ho	232 237		
		赤鳥	Chih Neaou	238 250		
		太元	Tae Yuen	251		
廢帝	Fei Te	神鳳	Shin Fung	252	31	221 252
		建興	Keen Hing	253		
		五鳳	Woo Fung	254 255		
		太平	Tae Ping	256 257	5	253 -257
景帝	King Te	永安	Yung Gan	258 263	6	258 -263
末帝	Mo Te	元興	Yuen Hing	264		
		甘露	Kan Loo	265		
		寶鼎	Paou Ting	266 268		
		建衡	Keen Hung	269 271		
		鳳皇	Fung Hwang	272 274		
		天册	Teen Tsih	275		
		天璽	Teen Se	276		
		天紀	Teen Ke	277 280	17	264 to 280

北魏 PIH, OR NORTHERN WEI.

Emperor's Name.		Epoch.		Duration of Epoch.	Reigned Years.	Duration of Reign.
道武帝	Taou Woo Te	登國	Tang Kwo	386 395		
		皇始	Hwang Che	396 to 397		

Emperor's Name.		Epoch.		Duration of Epoch.		Reigned Years.	Duration of Reign.	
				A. D.			A. D.	
		天興	Teen Hing	398 to 403				
		天賜	Teen Yang	404	408	23	386 to 408	
明元帝	Ming Yuen Te	永興	Yung Hing	409	413			
		神端	Shin Twan	414	415			
		泰常	Tae Chang	416	423	15	409	423
太武帝	Tae Woo Te	始光	Che Kwang	424	427			
		神麚	Shin Kea	428	431			
		延和	Yen Ho	432	434			
		太延	Tae Yen	435	439			
		太平	Tae Ping }	440	451			
		眞君	Chin Keun }					
		正平	Ching Ping	452		27	416	451
文成帝	Wan Ching Te	興安	Hing Gan	452	453			
		興光	Hing Kwang	454				
		太安	Tae Gan	455	459			
		和平	Ho Ping	460	465	14	452	465
獻文帝	Heen Wan Te	天安	Teen Gan	466				
		皇興	Hwang Hing	467	470	5	466	470
孝文帝	Heaou Wan Te	延興	Yen Hing	471	475			
		承明	Ching Ming	476				
		太和	Tae Ho	477	499	29	471	499
宣武帝	Seuen Woo Te	景明	King Ming	500	503			
		正始	Ching Che	504	507			
		永平	Yung Ping	508	511			
		延昌	Yen Chang	512 to 515		16	477 to 515	
孝明帝	Heaou Ming Te	熙平	He Ping	516				

Emperor's Name.			Epoch.		Duration of Epoch.		Reigned Years.	Duration of Reign.
					A.D.			A.D.
			神 龜	Shin Kwei	517 to 518			
			正 光	Ching Kwang	519	524		
			孝 昌	Heaou Chang	525	527	12	516 to 527
孝 莊 帝		Heaou Chwang Te	永 安	Yung Gan	528	530	3	528 530
東 海 王		Tung Hae Wang	建 明	Keen Ming	...	...	...	1 month
節 閔 帝		Tsee Min Te	晉 泰	Tsin Taè	531		1	531
安 定 王		Gan Ting Wang	中 興	Chung Hing	...	...	...	1 month
孝 武 帝		Heaou Woo Te	永 熙	Yung He	532	534	3	532 534

東 魏 TUNG, OR EASTERN WEI.

			Epoch.					
孝 靜 帝		Heaou Tsing Te	天 平	Teen Ping	534	537		
			元 象	Yuen Seang	538			
			興 利	Hing Le	539	542		
			武 定	Woo Ting	543	550	17	534 550

北 齊 PIH TSE, OR NORTHERN TSE.

			Epoch.					
文 宣 帝		Wan Seuen Te	天 保	Teen Paou	550	559	10	550 559
廢 帝		Fei Te	乾 明	Keen Ming	...	...	...	1 month
孝 昭 帝		Heaou Chaou Te	皇 建	Hwang Keen	560		1	560
武 成 帝		Woo Ching Te	大 寧	Ta Ning	561			
			河 清	Ho Tsing	562	564	4	561 564
後 主		How Choo	天 統	Teen Tung	565	569		
			武 平	Woo Ping	570 to 576		12	565 to 576
幼		Yew Choo	承 光	Ching Kwang	577		1	577

後 周 HOW CHOW, OR LATER CHOW. ALSO, PIH CHOW.

Emperor's Name.			Epoch.		Duration of Epoch.		Reigned Years.	Duration of Reign.	
					A. D.			A. D.	
明 帝	Ming Te	武 成		Woo Ching	557 to 560		4	557 to 560	
武 帝	Woo Te	保 定		Paou Ting	561	565			
		天 和		Teen Ho	566	571			
		建 德		Keen Tih	572	577			
		宣 政		Seuen Chang	578		18	561	578
宣 帝	Seuen Te	大 成		Ta Ching	A few months only.				
靜 帝	Tsing Te	大 象		Ta Seang	579	580			
		大 定		Ta Ting	581		3	579	581

遼 THE LEAOU, A TARTAR DYNASTY.

Emperor's Name.			Epoch.		Duration of Epoch.		Reigned Years.	Duration of Reign.	
太 祖	Tae Tsoo	No epoch for the first 9 years.		... ...	907	915			
		神 册		Shin Tsih	916	921			
		天 贊		Teen Tsan	922	925			
		天 顯		Teen Heen	926		20	907	926
太 宗	Tae Tsung	天 顯		Teen Heen	927	937			
		會 同		Hwuy Tung	938	946			
		大 同		Ta Tung	947		21	927	947
世 ‖	She Tsung	天 錄		Teen Luh	948	950	3	948	950
穆 ‖	Muh Tsung	應 歷		Ying Leih	951	968	18	951	968
景 ‖	King Tsung	保 寧		Paou Ning	969	978			
		乾 亨		Keen Hang	979	982	14	969	982
聖 ‖	Shing Tsung	統 和		Tung Ho	983	1011			
		開 泰		Kae Tae	1012	1020			
		太 平		Tae Ping	1021 to 1031		49	983 to 1031	

Emperor's Name.	Epoch.	Duration of Epoch.	Reigned Years.	Duration of Reign.
		A. D.		A. D.
興宗 Hing Tsung	景福 King Fuh	1032		
	重熙 Chung He	1033 to 1054	23	1032 to 1054
道宗 Taou Tsung	清寧 Tsing Ning	1055 1064		
	咸雍 Han Yung	1065 1074		
	太康 Tae Kang	1075 1084		
	大安 Ta Gan	1085 1094		
	壽隆 Show Lung	1095 1100	46	1055 1100
天作帝 Teen Tso Te	乾統 Keen Tung	1101 1110		
	天慶 Teen King	1111 1120		
	保大 Paou Ta	1121 1125	25	1101 1125

金 THE KIN, A TARTAR DYNASTY.

Emperor's Name.	Epoch.	Duration of Epoch.	Reigned Years.	Duration of Reign.
太祖 Tae Tsoo	天輔 Teen Foo	1118 1123	6	1118 1123
太宗 Tae Tsung	天會 Teen Hwuy	1124 1135	12	1124 1135
熙宗 He Tsung	天會 Teen Hwuy	1136 1139		
	天眷 Teen Keuen	1140 1142		
	皇統 Hwang Tung	1143 1151	16	1136 1151
海陵王 Hae Ling Wang	天德 Teen Tih	1152 1155		
	貞元 Ching Yuen	1156 1158		
	正隆 Ching Lung	1159 1163	12	1152 1163
世宗 She Tsung	大定 Ta Ting	1164 1192	29	1164 1192
章 King Tsung	明昌 Ming Chang	1193 1198		
	承安 Ching Gan	1199 1203		
	泰和 Tae Ho	1204 1211	19	1193 to 1211
衞紹王 Wei Shaou Wang	大安 Ta Gan	1212 to 1214		

Emperor's Name.		Epoch.		Duration of Epoch.		Reigned Years.	Duration of Reign.
				A. D.			A. D.
宣 宗	Seuen Tsung	崇 慶	Tsung King	1215			
		至 寧	Che Ning	1216		5	1212 to 1216
		貞 祐	Ching Yew	1217	1220		
		興 定	Hing Ting	1221	1226		
		元 光	Yuen Kwang	1227	1228	12	1217 1228
哀 |	Gae Tsung	正 大	Ching Ta	1229 to 1235			
		開 興	Kae Hing	1236			
		天 興	Teen Hing	1237		9	1229 to 1237

THE preceding Chronological Tables have been compiled from various historical works of repute. Among these it must be observed, that from the Tsin dynasty, B.C. 255, to the present time, the principal authorities which have been employed are the Japanese chronological work mentioned in the Introductory Remarks (p. xv.) and a series of eight Chinese rolls in the author's possession, which contain their chronology from the accession of the Tsin to the subversion of the Ming dynasty, A.D. 1644. As these rolls, in addition to the whole of the 'Neen Haou,' or epochs of the regular dynasties, record those of the principal minor dynasties, and as a collation with the 'She Ke' and other esteemed historical annals has proved them to be perfectly trustworthy, they form the chief authority for these epochs, their text being adopted throughout.

These Tables are to be employed for ascertaining the year of any historical or other event of which the date is required. In the early portion, the dates of the dynasty and emperor alone are mentioned, the 'Neen Haou,' or Epoch, not having been introduced until about 163 years before the Christian era. From that time, in addition to the above-mentioned dates, the year of the epoch is given; and this latter mode is that employed in the major part of the observations of comets in the treatises from which the present translation has been made. In the Chinese historical works, the mode of reckoning by cycles of 60 years is that usually followed.

In Table A. will be found the combinations of the Kea Tsze characters, by which the 60 years of these cycles are expressed; and Table C. shows the first year of each of them, from the first, commencing B.C. 2637, to the seventy-sixth, which began A.D. 1864. Table A. is also employed to express the periods of 60 days into which the Chinese year is divided, and whose appellations are the same as those of the years of the cycle. As this cycle of 60 years, although in constant use in the historical works, is not employed in expressing the dates of most of the cometary observations contained in the present publication, no mention of it occurs among the preceding examples of the reduction of Chinese time to our reckoning. This opportunity is therefore taken of explaining its use.

To find a given year of the cycle, and to express it in our manner, we must proceed as follows:—The date of the dynasty and of the accession of the Emperor having been ascertained from the Chronological Tables, the date of the first year of the cycle in which that Emperor flourished will appear from Table C, that of the first years of cycles. All that is then needed is to find in the 60-year Table A. the combination whose date is required, when the number above it will be that of the year of the cycle represented by that combination, and the corresponding year according to our reckoning can be easily ascertained. For example: in the 'Tung Keen Kang Muh' it is recorded, that during the reign of the Emperor Tae Tsung, of the Tang dynasty, in the year of the cycle 'Yih Maou,' an eclipse of the Sun occurred. On reference to the Chronological Tables, the date of the accession of this Emperor will be found to have been A.D. 763; which year Table C. shows to have fallen in the 57th cycle, whose first year was A.D. 724. In Table A. it will be seen that the combination 'Yih Maou' is the 52nd of the cycle, consequently the year required, according to our system, is A.D. 775.

As respects the ordinary use of these Chronological Tables, the instructions given in p. xvi. of the Introductory Remarks will be found amply sufficient.

Tables B. and D. are those required for finding the characters for the 1st of January in any year, B.C. or A.D. The first of these, B, contains the combinations of the Kea Tsze characters necessary to form the 80-year Table, whose construction is explained in the Introductory Remarks, p. xviii.; and D. is the auxiliary table, showing the first year of each period of 80 years, from B.C. 2561 to A.D. 2000, arranged under the letters B.C. and A.D.

Table E. shows the days on which the characters for January 1 recur, both in common and leap years. In Table F. will be found the first year of each lunar cycle of 19 years, from B.C. 609 to A.D. 1900; and Table G. gives the first day of each moon in every year of this cycle of 19 years. F. and G. must be considered as approximate only, but they are sufficiently accurate for the purpose required.

The Tables A, B, D, E, F, and G, are those to be employed in finding the moons and days, and as their use is fully explained in the Introductory Remarks, pp. xv.–xx., they need no further notice here.

In the Plate marked H will be found the Tables referred to in pp. xxii. and xxiii. of the Introductory Remarks, the first being that of the Tsze Ke, or twenty-four divisions of the year, and the second that of the twelve Kung; and, it may be observed, it would appear that the names of these latter, not being anywhere described as referring to existing asterisms, as composing them, are to be considered as indicating divisions only, rather than individual groups of stars. It must also be remarked that the modern names, as far as at present has been ascertained, do not occur in any astronomical treatise whose compilation dates before the accession of the present dynasty.

癸	壬	辛	庚	巳	戊	丁	丙	乙	甲
Kwei	Jin	Sin	Kang	Ke	Woo	Ting	Ping	Yih	Kea

亥	戌	酉	申	未	午	巳	辰	卯	寅	丑	子
Hae	Seuh	Yew	Shin	wei	woo	sze	Shin	maou	yin	chow	Tsze

A.

The Combinations of the Kea Tsze Characters forming the Cycle of sixty years or the periods of sixty days

57 寅甲 Kea yin	41 辰甲 Kea Shin	31 午甲 Kea Woo	21 申甲 Kea Shin	11 戌甲 Kea Seuh	1 子甲 Kea Tsze
52 卯乙 Yih maou	42 巳乙 Yih Sze	32 未乙 Yih Wei	22 酉乙 Yih Yew	12 亥乙 Yih Hae	2 丑乙 Yih Chow
53 辰丙 Ping Shin	43 午丙 Ping Woo	33 申丙 Ping Shin	23 戌丙 Ping Seuh	13 子丙 Ping Tsze	3 寅丙 Ping Yin
54 巳丁 Ting Sze	44 未丁 Ting Wei	34 酉丁 Ting Yin	24 亥丁 Ting Hae	14 丑丁 Ting Chow	4 卯丁 Ting maou
55 午戊 Woo Woo	45 申戊 Woo Shin	35 戌戊 Woo Seuh	25 子戊 Woo Tsze	15 寅戊 Woo Yin	5 辰戊 Woo Shin
56 未巳 Ke Wei	46 酉巳 Ke Yew	36 亥巳 Ke Hae	26 丑巳 Ke Chow	16 卯巳 Ke maou	6 巳巳 Ke sze
57 申庚 Kang Shin	47 戌庚 Kang seuh	37 子庚 Kang Tsze	27 寅庚 Kang Yin	17 辰庚 Kang Shin	7 午庚 Kang Woo
58 酉辛 Sin yew	48 亥辛 Sin Hae	38 丑辛 Sin chow	28 卯辛 Sin maou	18 巳辛 Sin sze	8 未辛 Sin Wei
59 戌壬 Sin Seuh	49 子壬 Sin Tsze	39 寅壬 Sin Yin	29 辰壬 Sin Shin	19 午壬 Sin Woo	9 申壬 Sin Shin
60 亥癸 Kwei Hae	50 丑癸 Kwei Chow	40 卯癸 Kwei maou	30 巳癸 Kwei sze	20 未癸 Kwei Wei	10 酉癸 Kwei Yew

B.

No.	Char.		No.	Char.		No.	Char.		No.	Char.	
1	丑丁 [14]	Ting chow	21	戌壬 [59]	Jin Seuh	41	未丁 [44]	Ting Wei	61	辰壬 [29]	Jin Shin
2	午壬 [19]	Jin Woo	22	卯丁 [4]	Ting maou	42	子壬 [49]	Jin Tsze	62	酉丁 [34]	Ting Yew
3	亥丁 [24]	Ting Hae	23	申壬 [9]	Jin Shin	43	巳丁 [54]	Ting sze	63	寅壬 [39]	Jin Yin
4	辰壬 [29]	Jin Shin	24	丑丁 [14]	Ting Chow	44	戌壬 [59]	Jin seuh	64	未丁 [44]	Ting Wei
5	戌戊 [35]	Woo Seuh	25	未癸 [20]	Kwei Wei	45	辰戊 [5]	Woo Shin	65	丑癸 [50]	Kwei Chow
6	卯癸 [40]	Kwei maou	26	子戊 [25]	Woo Tsze	46	酉癸 [10]	Kwei yew	66	午戊 [55]	Woo Woo
7	申戊 [45]	Woo Shin	27	巳癸 [30]	Kwei sze	47	寅戊 [15]	Woo Yin	67	亥癸 [60]	Kwei Hae
8	丑癸 [50]	Kwei Chow	28	戌戊 [35]	Woo Seuh	48	未癸 [20]	Kwei Wei	68	辰戊 [5]	Woo Shin
9	未巳 [56]	Ke Wei	29	辰甲 [41]	Kea Shin	49	丑巳 [26]	Ke Chow	69	戌甲 [11]	Kea Seuh
10	子甲 [1]	Kea Tsze	30	酉巳 [46]	Ke yew	50	午甲 [31]	Kea Woo	70	卯巳 [16]	Ke maou
11	巳巳 [6]	Ke Sze	31	寅甲 [51]	Kea Yin	51	亥巳 [36]	Ke Hae	71	申甲 [21]	Kea Shin
12	戌甲 [11]	Kea Seuh	32	未巳 [56]	Ke Wei	52	辰甲 [41]	Kea Shin	72	丑巳 [26]	Ke Chow
13	辰庚 [17]	Kang Shin	33	丑乙 [2]	yih Chow	53	戌庚 [47]	Kang Seuh	73	未乙 [32]	yih Wei
14	酉乙 [22]	yih yew	34	午庚 [7]	Kang Woo	54	卯乙 [52]	yih maou	74	子庚 [37]	Kang Tsze
15	寅庚 [27]	Kang Yin	35	亥乙 [12]	yih Hae	55	申庚 [57]	Kang Shin	75	巳乙 [42]	yih sze
16	未乙 [32]	yih wei	36	辰庚 [17]	Kang Shin	56	丑乙 [2]	yih Chow	76	戌庚 [47]	Kang Seuh
17	丑辛 [38]	Sin Chow	37	戌丙 [23]	Ping Seuh	57	未辛 [8]	Sin Wei	77	辰丙 [53]	Ping Shin
18	午丙 [43]	Ping Woo	38	卯辛 [28]	Sin maou	58	子丙 [13]	Ping Tsze	78	酉辛 [58]	Sin yew
19	亥辛 [48]	Sin Hae	39	申丙 [33]	Ping Shin	59	巳辛 [18]	Sin Sze	79	寅丙 [3]	Ping Yin
20	辰丙 [53]	Ping Shin	40	丑辛 [38]	Sin Chow	60	戌丙 [23]	Ping Seuh	80	未辛 [8]	Sin wei
									81	丑丁 [14]	Ting Chow

The 80 year Table for finding the Characters for the 1st of January in any year B.C. or A.D.

C.

The first year of each Cycle of 60 years, from BC 2637 to AD 1864

I	BC 2637	XIV	1857	XXVII	1077	XL	297	LIII	484	LXVI	1264
II	2577	XV	1797	XXVIII	1017	XLI	237	LIV	544	LXVII	1324
III	2517	XVI	1737	XXIX	957	XLII	177	LV	604	LXVIII	1384
IV	2457	XVII	1677	XXX	897	XLIII	117	LVI	664	LXIX	1444
V	2397	XVIII	1617	XXXI	837	XLIV	57	LVII	724	LXX	1504
VI	2337	XIX	1587	XXXII	777	XLV	AD 4	LVIII	784	LXXI	1564
VII	2277	XX	1497	XXXIII	717	XLVI	64	LIX	833	LXXII	1624
VIII	2217	XXI	1437	XXXIV	657	XLVII	124	LX	904	LXXIII	1684
IX	2157	XXII	1377	XXXV	597	XLVIII	184	LXI	964	LXXIV	1744
X	2097	XXIII	1317	XXXVI	537	XLIX	244	LXII	1024	LXXV	1804
XI	2037	XXIV	1257	XXXVII	477	L	304	LXIII	1084	LXXVI	1864
XII	1977	XXV	1197	XXXVIII	417	LI	364	LXIV	1144		
XIII	1917	XXVI	1137	XXXIX	357	LII	424	LXV	1204		

D

The commencement of each period of 80 years from BC 2561

BC 2561	2081	1681	1281	881	481	81	AD 80	480	880	1280	1680
2401	2001	1601	1201	801	401	1	160	560	960	1360	1760
2321	1921	1521	1121	721	321		240	640	1040	1440	1840
2241	1841	1441	1041	641	241		320	720	1120	1520	1920
2161	1761	1361	961	561	161		400	800	1200	1600	2000

E

The days on which the characters for January 1st recur

Common Years		Leap years	
March	2	March	1
May	1	April	30
June	30	June	29
August	29	August	28
October	28	October	27
December	27	December	26

F.

The first year of each Cycle of 19 years from B.C. 609 to A.D. 1900

BC	419	229	39	133	325	513	703	893	1083	1273	1463	1653	1843
609	400	210	20	152	342	532	722	912	1102	1292	1482	1672	1862
590	381	191	1	171	361	551	741	931	1121	1311	1501	1691	1881
571	362	172	AD	190	380	570	760	950	1140	1330	1520	1710	1900
552 / 533	343	153	19	209	399	589	779	969	1159	1349	1539	1729	
514	324	134	38	228	418	608	798	988	1178	1368	1558	1748	
495	305	115	57	247	437	627	817	1007	1197	1387	1577	1767	
476	286	96	76	266	456	646	836	1026	1216	1406	1596	1786	
457	267	77	95	285	475	665	855	1045	1235	1425	1615	1805	
438	248	58	114	304	494	684	874	1064	1254	1444	1634	1824	

G.

Approximate Table of the first day of each Moon for every year of the Lunar Cycle of 19 years

	Jan	Feb	Mar	Apr	May	June	July	Aug	Sept	Oct	Nov	Dec	Dec
1	Jan 23	Feb 21	Mar 23	Apr 21	May 21	June 13	July 19	Aug 17	Sept 16	Oct 15	Nov 14	Dec 13	
2	Jan 12	Feb 10	Mar 12	Apr 10	May 10	June 8	July 8	Aug 6	Sept 5	Oct 2	Nov 3	Dec 2	
3	Jan 1	Jan 31	Mar 1	Mar 31	Apr 29	May 29	June 27	July 27	Aug 25	Sep 24	Oct 23	Nov 22	Dec 21
4	Jan 20	Feb 18	Mar 20	Apr 18	May 18	June 16	July 16	Aug 14	Sep 13	Oct 12	Nov 11	Dec 10	
5	Jan 9	Feb 7	Mar 9	Apr 7	May 7	June 5	July 5	Aug 3	Sept 2	Oct 2	Oct 31	Nov 30	Dec 29
6	Jan 28	Feb 26	Mar 28	Apr 26	May 26	June 24	July 24	Aug 22	Sept 21	Oct 20	Nov 19	Dec 18	
7	Jan 17	Feb 15	Mar 17	Apr 15	May 15	June 13	July 12	Aug 11	Sept 10	Oct 9	Nov 8	Dec 7	
8	Jan 6	Feb 4	Mar 6	Apr 5	May 4	June 3	July 2	Aug 1	Aug 30	Sep 29	Oct 28	Nov 27	Dec 26
9	Jan 25	Feb 23	Mar 25	Apr 23	May 23	June 21	July 21	Aug 19	Sept 18	Oct 17	Nov 16	Dec 16	
10	Jan 14	Feb 12	Mar 14	Apr 13	May 12	June 10	July 10	Aug 8	Sep 7	Oct 6	Nov 5	Dec 4	
11	Jan 3	Feb 2	Mar 3	Apr 2	May 1	May 31	June 29	July 29	Aug 27	Sept 26	Oct 25	Nov 24	Dec 23
12	Jan 22	Feb 20	Mar 22	Apr 20	May 20	June 18	July 18	Aug 16	Sep 15	Oct 14	Nov 13	Dec 12	
13	Jan 11	Feb 9	Mar 11	Apr 9	May 9	June 7	July 7	Aug 5	Sept 4	Oct 3	Nov 2	Dec 1	Dec 31
14	Jan 30	Feb 28	Mar 30	Apr 28	May 28	June 26	July 26	Aug 24	Sep 23	Oct 22	Nov 21	Dec 20	
15	Jan 19	Feb 17	Mar 19	Apr 17	May 17	June 15	July 15	Aug 13	Sept 12	Oct 11	Nov 10	Dec 9	
16	Jan 8	Feb 6	Mar 8	Apr 6	May 6	June 4	July 4	Aug 2	Sept 1	Oct 1	Oct 30	Nov 29	Dec 28
17	Jan 27	Feb 25	Mar 27	Apr 25	May 25	June 23	July 23	Aug 21	Sep 20	Oct 19	Nov 18	Dec 17	
18	Jan 16	Feb 14	Mar 16	Apr 14	May 14	June 12	July 12	Aug 10	Sept 9	Oct 8	Nov 7	Dec 6	
19	Jan 5	Feb 3	Mar 5	Apr 4	May 3	June 2	July 1	July 30	Aug 28	Sep 29	Oct 26	Nov 25	Dec 24

H.

The Twenty four divisions of the Chinese Year

Chinese	Romanization	Meaning	Date
小寒	Seaou Han	moderate Cold	Jan 6
大寒	Ta Han	Great Cold	" 21
立春	Leih Chun	Spring begins	Feb 5
雨水	Yu Shwuy	Rain Water	" 21
驚蟄	King chih	Worms move	Mar 6
春分	Chun Fun	Spring middle	Mar 22
清明	Tsing ming	Pure brightness	Apr 6
穀雨	Kuh yu	Grain rain	" 22
立夏	Leih Hea	Summer begins	May 7
小滿	Seaou mwan	a little full	" 22
芒種	Mang chung	Grain in the ear	June 7
夏至	Hea che	Summer Height	" 22

Chinese	Romanization	Meaning	Date
小暑	Seaou Shoo	Moderate Heat	July 8
大暑	Ta Shoo	Great Heat	" 21
立秋	Leih Tsew	autumn begins	Aug 9
處暑	Choo Shoo	Later Heat	" 24
白露	Pih Loo	White Dew	Sep 9
秋分	Tsew Fun	autumn middle	" 22
寒露	Han Loo	Cold Dew	Octr 9
霜降	Shwang Keang	Hoar Frost	" 21
立冬	Leih Tung	winter begins	Nov 8
小雪	Seaou Seuh	moderate snow	" 23
大雪	Ta Seuh	Great Snow	Dec 8
冬至	Tung Che	Winter middle	" 22

The Twelve Kung or divisions of the Ecliptic answering to our Zodiacal Signs

Ancient — 中國名 Chung kwo ming — Chinese names

Modern — 西國名 Se kwo ming — European names

Chinese names	Romanization	Translation	European names	Romanization	Meaning
娵訾	Heang low		白羊	Pih Yang	The White Sheep
降婁	Ta Leang	The Great Bridge	金牛	Kin New	The Golden Ox
大梁	Shih chin		陰陽	Yin Yang	The two Principles
實沈	Shun Show	The Quail's Head	巨蟹	Keu Hea	The Crab
鶉首	Shun Ho	The Quail's Fire	獅子	Sze Tsze	The Lion
鶉火	Shun Wei	The Quail's Tail	霜女	Shwang New	The Frigid maiden
鶉尾	Show Sing	The Star of Longevity	天秤	Teen ching	The Celestial Balance
壽星	Ta Ho	The Great Fire	天蝎	Teen Hee	The Celestial Scorpion
大火	Seih muh	The Cleft Tree	人馬	Jin ma	The man Horse
析木	Sing Ke	Starry record	磨羯	Mo Ke	A kind of Sheep or Goat!
星紀	Yuen Heaou		寶瓶	Paou King	The precious Vase
元枵	Tseu Tsze		雙魚	Shwang Yu	The two Fishes

The translation of some of the ancient names is so unsatisfactory that it has been thought advisable to omit it. Morrison in his dictionary refers to them simply as the names of certain stars or Constellations without attempting to identify them.

Plate 1.

CHINESE CELESTIAL ATLAS

The Twenty Eight Stellar Divisions their determining Asterisms, their Extent North and South and East and west, with the three great Spaces.

No	Names		Determining Asterisms	Extent N & S.	E & W on the Equator
1	角	Keo	α and ζ Virginis	Coma Berenices to Centaurus	11°.49
2	亢	Kang	ι κ λ φ Virginis	Bootes to Lupus	9.19
3	氐	Te	α β γ ι Libræ	Bootes to Lupus	16.41
4	房	Fang	β δ π ρ Scorpii	Ophiuchus to Lupus	5.28
5	心	Sin	α σ τ Scorpii	These and stars near	6.9
6	尾	Wei	ε μ ν Scorpii	Id Id	21.6
7	箕	Ke	ν δ ε θ Sagittarii	Id Id	8.46
8	斗	Tow	ζ τ σ φ λ μ Sagittarii	Id Id	24.24
9	牛	new	α β &c. Capricorni	Lyra to Capricornus	6.50
10	女	neu	ε μ ν &c. Aquarii	Cygnus to Aquarius	11.7.
11	虚	Heu	β Aquarii and another	Equuleus to Grus	8.41
12	危	Wei	α Aquarii θ.ε. Pegasi	Cepheus to Piscis Notius	14.53
13	室	Shih	α β Pegasi &c.	Cygnus to Id.	17.0
14	壁	Peih	γ Pegasi α Andromeda	Pegasus to Cetus	10.28
15	奎	Kwei	β δ &c Andromeda & stars in Pisces	Cassiopeia to Cetus	14.30
16	婁	Lew	α β γ Arietis	Andromeda to Cetus	12.4
17	胃	wei	The 3 stars in musca	Perseus to Cetus	15.45
18	昴	maou	The Pleiades	Perseus to Eridanus	10.24
19	畢	Pech	α γ δ ε &c Tauri	Auriga to Eridanus	16.34
20	觜	Tsuy	λ and others in Orions head	small stars near	.24
21	參	Tsan	α β γ δ &c Orionis	Orion to Lepus	11.24
22	井	Tsing	γ ε λ μ &c. Geminorum	Perseus to argo	32.49
23	鬼	Kwei	γ δ η θ Cancri	Cancer to argo	2.21
24	柳	Lew	δ ε ζ θ Hydræ	leo to Hydra	12.14
25	星	Sing	α τ &c. Hydræ	Leo minor to Hydra	5.48
26	張	chang	κ λ μ &c. Hydræ	Ursa Major to Hydra	17.19
27	翼	Yih	α and others in Crater	a few stars near	20.25
28	軫	chin	β &c Corvi	Id Id	15.30

紫微垣	Tsze Wei queen	The northern circumpolar stars
天市垣	Tsen She yuen	Space bounded by serpens
太微垣	Tae Wei Yuen	Space within stars on leo and Virgo

Plate 2.

紫 Tze
微 Wei
垣 Yuen

The Northern Circumpolar Stars
as many of the Groups cannot
be identified their position alone is given

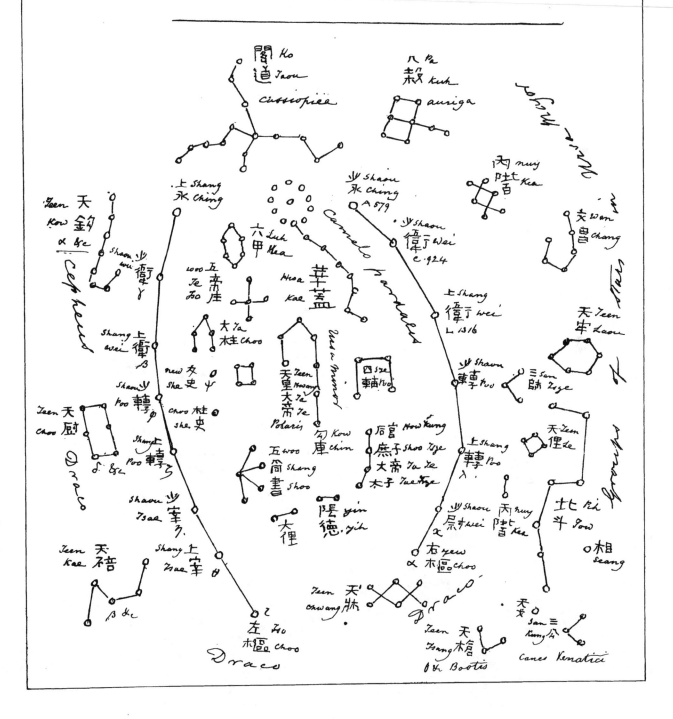

Plate 3.

天 Teen
市 She
垣 Yuen

The Space bounded by. Serpens comprising Hercules The upper part of Ophiuchus &c

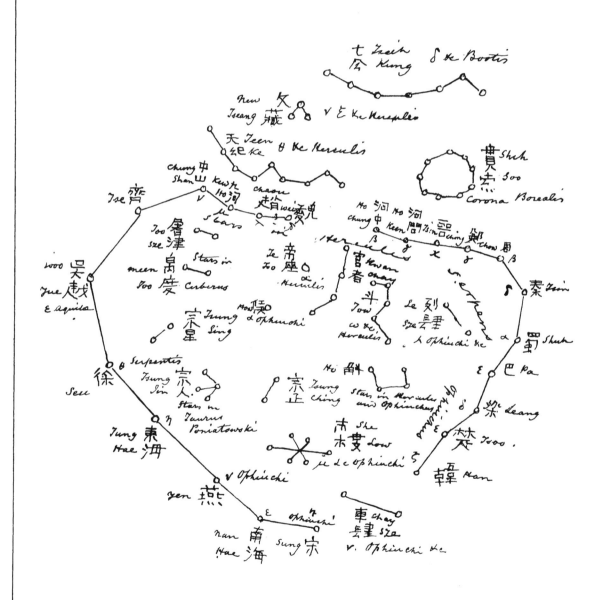

Plate 4.

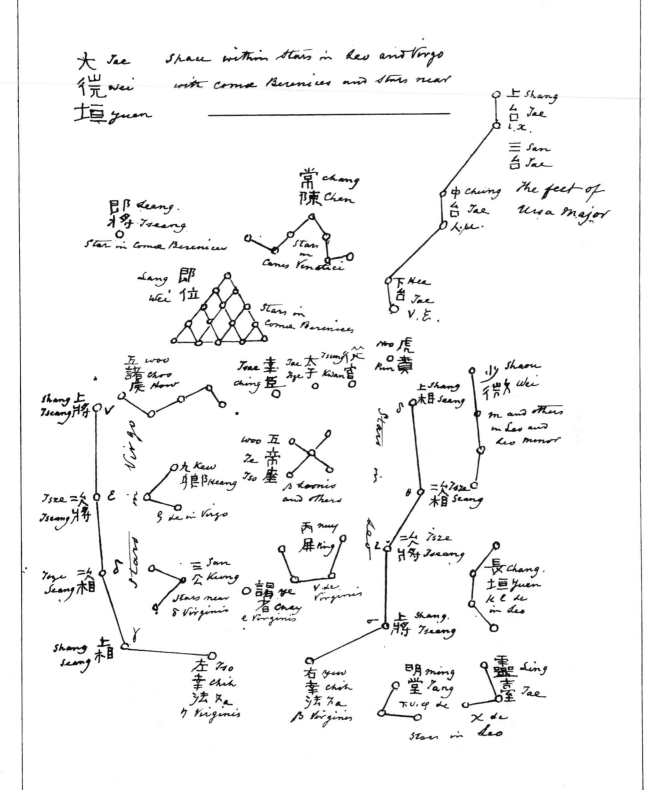

大 Tae
微 wei
垣 yuen

Space within stars in Leo and Virgo
with coma Berenices and stars near

上 Shang
台 Tae
上.太.

三 San
台 Tae

中 chung
台 Tae
中.太.

The feet of
Ursa Major

常 chang
陳 Chen
Stars in
Canes Venatici

郎 Leang.
將 Tseang
Star in Coma Berenices

下 Hea
台 Tae
V. ε.

郎 Lang
位 wei
Stars in
Coma Berenices

五 woo
諸 choo
虎 How

宰 ching
相

太 Tae
子 Tsze
宗 Tsung
官 Kwan

虎 Hoo
賁 Pun

上 Shang
相 Seang

少 Shaou
微 Wei
m and others
in Leo and
Leo minor

Shang 上
Tseang 將
V.

在 Virgo
in

五 woo
帝 Te
座 Tso
β Leonis
and others

二 Tsze
相 Seang

Tsze 二
Tseang 將
ε

九 Kew
娘 Keang
g de in Virgo

丙 ney
屏 Ping

Stars in
Leo

二 Tsze
將 Tseang

長 chang.
垣 Yuen
k l de
in Leo

Tsze 二
Seang 相
δ

三 San
公 Kung
Stars near
δ Virginis

謁 yĕ
者 Chay
ε Virginis

V. de
Virginis

上 Shang.
將 Tseang

Shang 上
Seang 相
γ

左 Tso
執 chih
法 Fa
η Virginis

右 yew
執 chih
法 Fa
β Virginis

明 ming
堂 Tang
T. U. q de

靈 Ling
臺 Tae
X de

Stars in Leo

The Twenty Eight Stellar Divisions.

Plate 5.

1 角 Keo

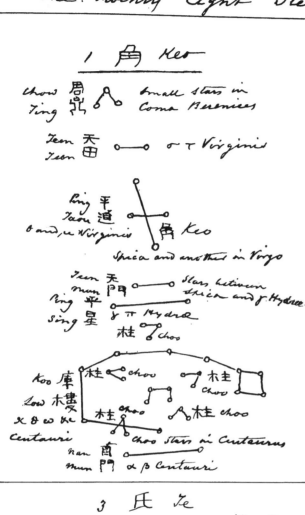

Chow Ting 爵鼎 — Small stars in Coma Berenices

Teen Teen 天田 — σ τ Virginis

Ping Taou 平道 — Keo 角
δ and η Virginis
Spica and another in Virgo

Teen mun 天門
Ping 平
Sing 星 — Stars between Spica and ? Hydrae
ξ π Hydrae
柱 choo

Koo Low 庫樓
x θ ω &c Centauri
柱 choo
柱 choo
柱 choo
choo stars in Centaurus

Nan mun 南門 — α β Centauri

2 六 Kang

Yew She Tso 右攝提
η τ ν Bootes

大角 Ta Keo
Arcturus

Tso She Tso 左攝提
ζ π &c Bootes

六 Kang — ι κ λ θ Virginis

折 che 威 wei
Stars in Centaurus and Lupus

Teen Hang 天頓頑真
陽門 Yang mun

3 氐 Te

垣 Chaou
擺 Taou
β Bootis

Kang Ho 梗河
ε ς σ Bootis

Star near ε Serpentis
天乳 Teen Joo
帝座 Te Seih

α β γ ι Libra
氐 Te

Kang che 亢池 亢
Stars in Bootes

Stars in Scorpio
天輪 Teen Juh
陣車 Chin Chay
3 Stars in Scorpio
騎官 Ke Kwan

陳將軍 Chin Tseang Keun
S Lupi
27 Stars in Lupus in Threes

車 Chay
騎 Ke
π λ μ Lupi

4 房 Zang.

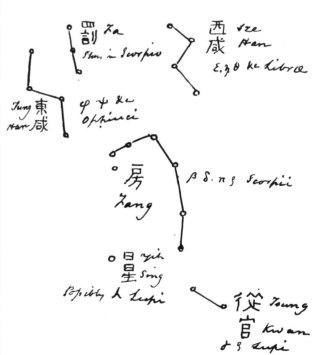

罰 Za
Stars in Scorpio

西咸 Sze Han
ε 3 θ &c Libra

Tung Han 東咸
φ ψ &c Ophiuci

房 Zang
β δ π ς Scorpii

日 yih 星 Sing
Possibly ι Lupi

從官 Tsung Kwan
θ ς Lupi

Plate 6.

5 心 Sin

心 Sin

Antares and σ τ Scorpii

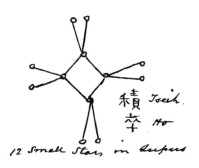

積 Tseih
卒 Tso

12 Small Stars in Lupus

6 尾 Wei

天 Teen
江 Keang

θ and others in Ophiuchus
and Sagittarius

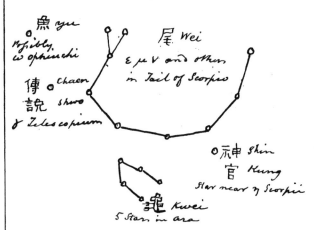

魚 yu

possibly,
ω ophiuchi

傅 chuen
說 shwo
θ Telescopium

尾 Wei

ε μ ν and others
in Tail of Scorpio

示申 Shin
官 Kwan
Star near η Scorpii

龜 Kwei
5 Stars in ara

7 箕 Ke

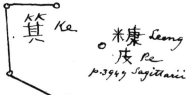

箕 Ke

米康 Keang
庚 Pe
p.394? Sagittarii

γ. S. E. Sagittarii
and β Telescopium

杵 much
杵 choo

α and others in ara

8 斗 Tow

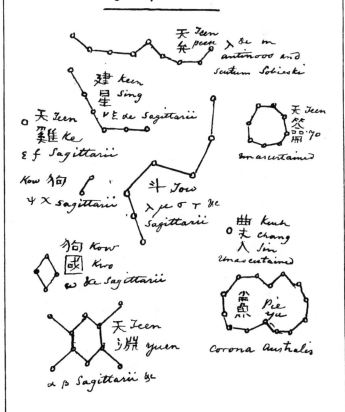

天 Teen
弁 peen

λ &c in
antinoos and
scutum Sobieski

建 keen
星 Sing
ν ε &c Sagittarii

天 Teen
雞 Ke
ε f Sagittarii

Kow 狗
4 X Sagittarii

斗 Tow
λ μ σ τ &c
Sagittarii

天 Teen
笠 Teen
70
unascertained

曲 keuh
天 chang
入 Sin
unascertained

狗 Kow
國 Kwo
ω &c Sagittarii

小黑 Pie
魚 yu
corona Australis

天 Teen
淵 yuen

α β Sagittarii &c

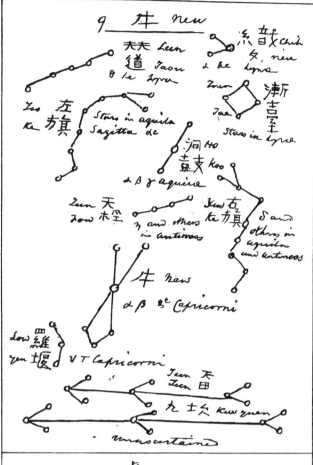
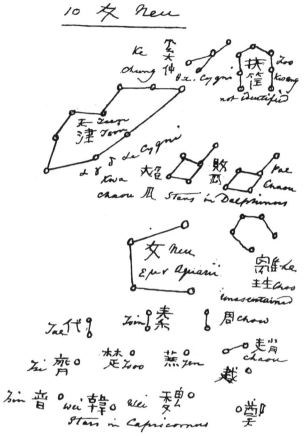
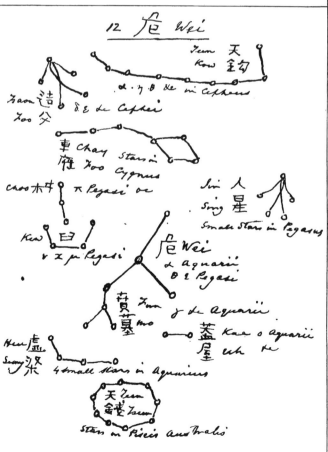

Plate 7.

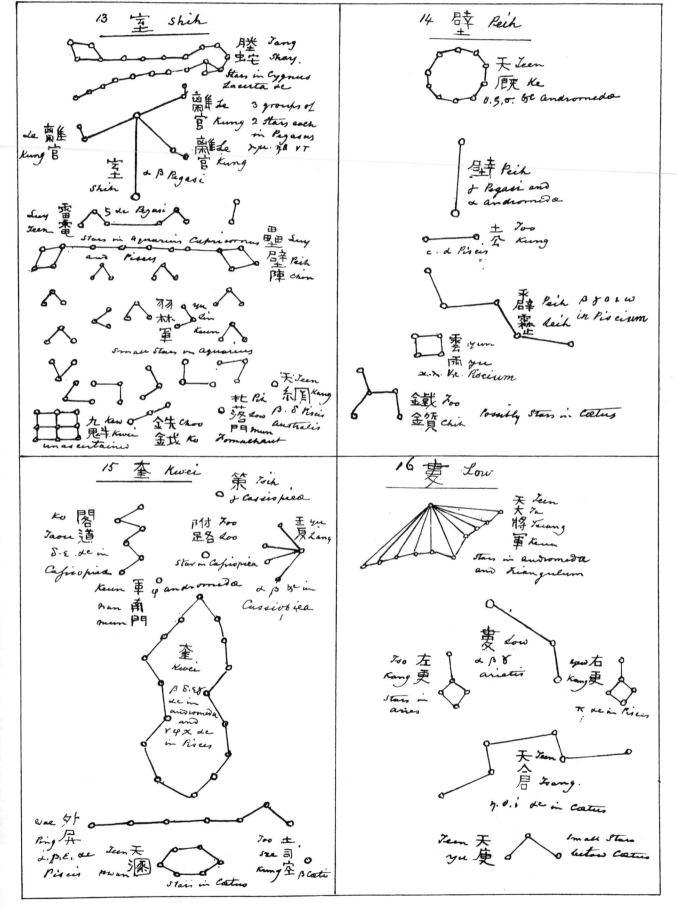

Plate 8.

Plate 9.

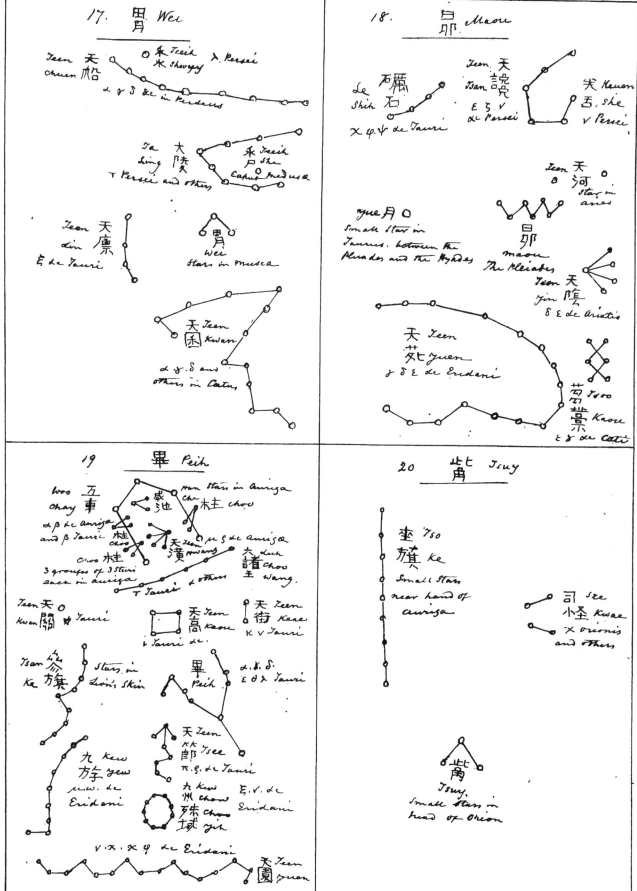

Plate 10.

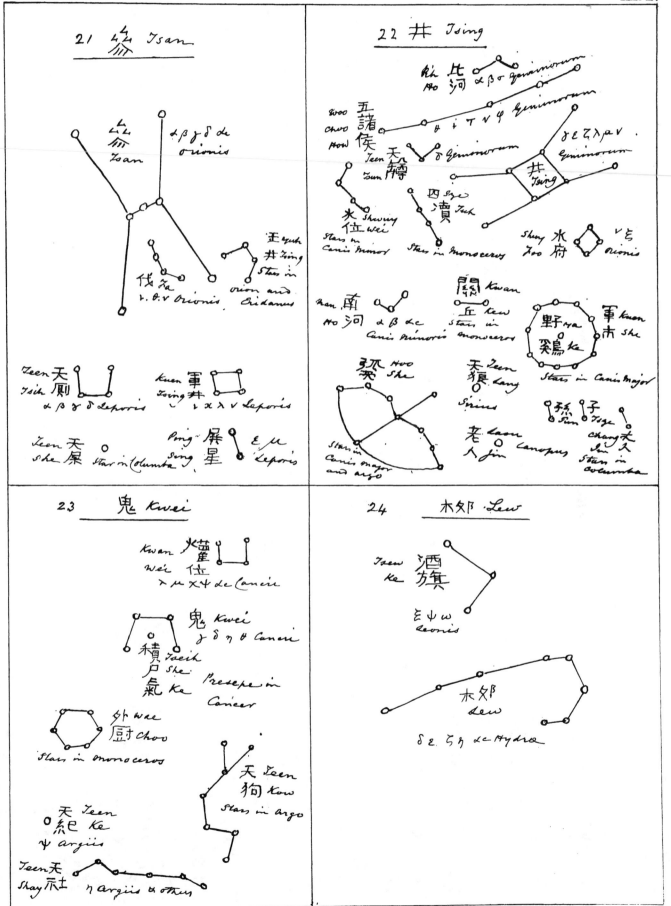

Plate II.

25 星 Sing

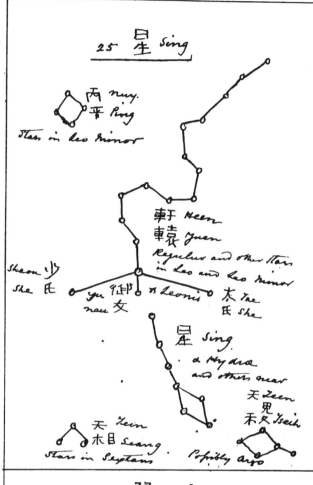

丙 muy. 平 Ping
Stars in Leo minor

軒 Heen 轅 Yuen
Regulus and other stars in Leo and Leo minor

Shaou 少 She 氏

御 yu 女 neu

α Leonis

太 Tae 氏 she

星 Sing
α Hydræ and others near

天 Teen 兒尺 Tseih

天 Teen 木目 Seang
Stars in Sextans

Possibly Argo

26 張 Chang

少 Shaou 微 Wei
m &c in Leo

天 Teen 尊 Tsun
ψ Ursæ majoris

長 Chang. 垣 Hwan
k l &c in Leo

張 Chang.
α λ μ &c, Hydræ

天 Teen 廟 meaou
Stars in Argo

27 翼 Yih

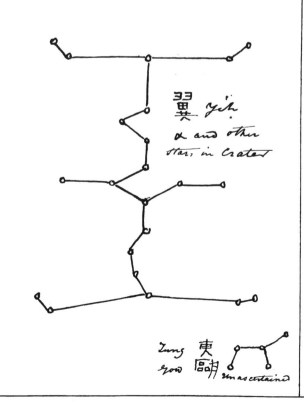

翼 Yih
α and other stars in Crater

東 Tung 甌 ?
unascertained

28 軫 Chin

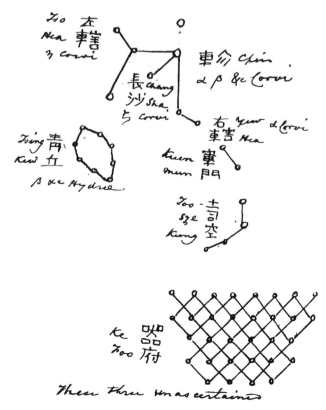

左 Tso 轄 Hea
3 Corvi

長 Chang 沙 Sha
5 Corvi

軫 Chin
α β &c Corvi

右 Yew 轄 Hea
d Corvi

青 Tsing 丘 Kew
β &c Hydræ

軍 kuen 門 mun

土 Szè 司 空 Kung

器 Ke 府 Foo

These three unascertained

Plate 12.

The Southern Circumpolar Stars

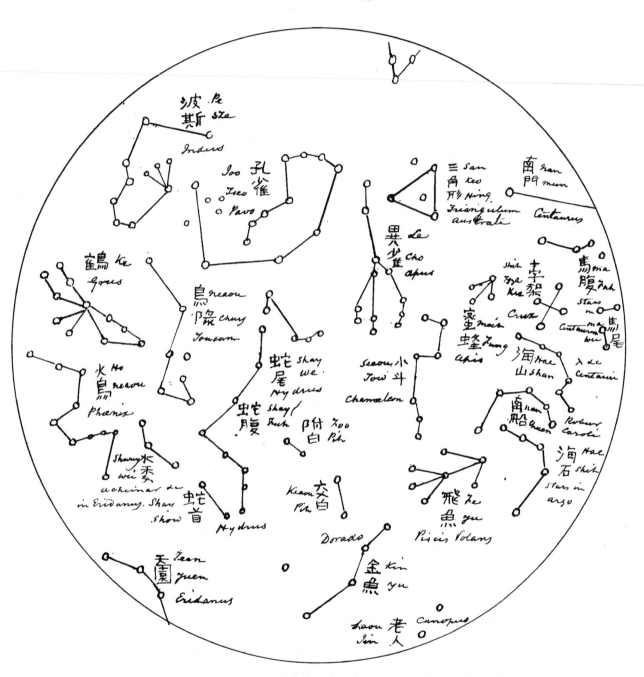

Name			Name			Name			Name		
Shang Wei	2		Teen chuen	9	17	Teen Tseen	7	12	Tsze	10 (10)	22
Shaou ching	2		Teen chwang	2		Teen Tsih	10	21	Tsze Tsang	4	
Shaou Poo	2		Teen Tow	7	9	Teen yuen	6	8	Tsze Tsang	4	
Shaou she	11	25	Teen Tuh	5	3	Teen yuen	9	18	Tung Yow	11	27
Shaou Tsae	2		Teen Hang	5	2	Teen yuen	9	19	Tung Hae	3	
Shaou wei	2		Teen Hwan	8	15	Too Kung	8	14	Tung Han	5	4
Shaou wei	4		Teen Hwang	9	19	Too Sze —	3		b		
Shaou Wei	11	26	Teen Hwang Ta Te	2		Too Sze kung	8	15	Wae choo	10	23
Shay Tuh	12		Teen Loo	5	3	Too Sze Kung	11	28	Wae King	8	15
Shay Show	12		Teen Tae	2		Tow	3		Wan chang	2	
Shay we	12		Teen Kang	8	13	Tow S.D.	6	8	Wei	3	
She low	3		Teen Kaou	9	19	Tsanke	9	19	Wei	7	10
Shih S.D	8	13	Teen Ke	3		Tsan. S.D	10	21	Wei	7	10
Shih Tze Kea	12		Teenke or ke	6	8	Tsan Tae	7	9	Wei S D	6	6
Shin Kung	6	6	Teen Ke	10	23	Tsae Ching	4		woo chuy	9	19
Shoo Tze	2		Teen Keae	9	19	Tsaou foo	7	12	woo choo Kow	4	
Shuh	3		Teen Keang	6	6	Tse	3		woo choo Kow	10	22
Shury Zoo	10	22	Teen Kow	2		Tse Tseang	4		woo Shang Shoo	2	
Shury Wei	12		Teen Kow	10	23	Tseih Ho	6	5	Woo Te Tso	2	
Shury Wei	10	22	Teen Kow	7	12	Tseih Kung	4		Woo Te Tso	4	
Sin S.D	6	5	Teen Kwan	9	17	Tseih Ho	6	5	woo yue	3	
Sing S.D	11	25	Teen Kwan	9	19	Tseih Kung	3		Wei S.D	7	12
Shih Loo	3		Teen Lang	10	22	Tseih She	9	17	Wei S.D	9	17
Sun	10	22	Teen Laou	2		Tseih Sheke	10	23			
Sung	3		Tun Le	2		Tseih Shury	9	17	Ya Ke	10	22
Sze Tsei	7	11	Teen lin	9	17	Tsew Ke	10	24	yang mun	5	2
Sze Ke	7	11	Teen luy Ching	7	11	Tsi	7	10	yen	3	
Sze Kwei	9	20	Teen meaou	11	26	Tsih	8	15	yen	7	10
Sze Poo	2		Teen mun	5	1	Tsin	3		yuw chih fa	4	
Sze Tuh	10	22	Teen 0 or Ho	9	18	Tsun	7	10	yuw choo	2	
Sze Wei	7	11	Teen peen	6	8	Tsing Kew	11	28	yuw Hea	11	28
Sze ming	7	11	Teen Seang	11	25	Tsung S.D	10	22	yuw Kung	8	6
			Teen Shay	11	23	Tso choo	2		yuw ke	7	9
Ta choo	2		Teen She	10	21	Tso chih Ta	4		yuw She Ta —	5	2
Ta Keo	5	1	Teen Ta Tseang Keun	8	16	Tso Hea	11	28	Tye chay	4	
Ta Lee	2		Teen Teen	5	1	Tso Kang	8	16	yih Sing	5	4
Ta Ling	9	17	Teen Tsan	9	18	Tso Ke	7	9	yih S D	11	27
Ta Te	2		Teen Tsang	2		Tso Ke	9	20	yin yih	2	
Tae	7	10	Teen Tsee	9	19	Tso she Ta	5	2	yu	6	6
Tae she	11	25	Teen Tseh	11	25	Tsoo	3		Yu Lin keun	8	13
Tae Tsze	2		Teen Tsin	7	10	Tsoo	7	10	yu nen	11	25
Tae Tsze	4		Teen Tsun	10	22	Tsoo Kaou	9	18	yue	9	18
Tang Shay	8	13	Teen Tsun	11	26	Tsung Tin	3		yuh Lang	8	15
Te Zoo	3		Teen yih	2		Tsung Tsing or chin	3		yuh Tsing	10	21
Te leih	5	3	Teen yn	9	18	Tsung Kwan	5	24	yun Tse	8	14
Te S.D	5	3	Teen yo	6	8	Tsung Kwan	4				
Teen choo	2		Teen yu	8	16	Tsuy S.D	9	30			

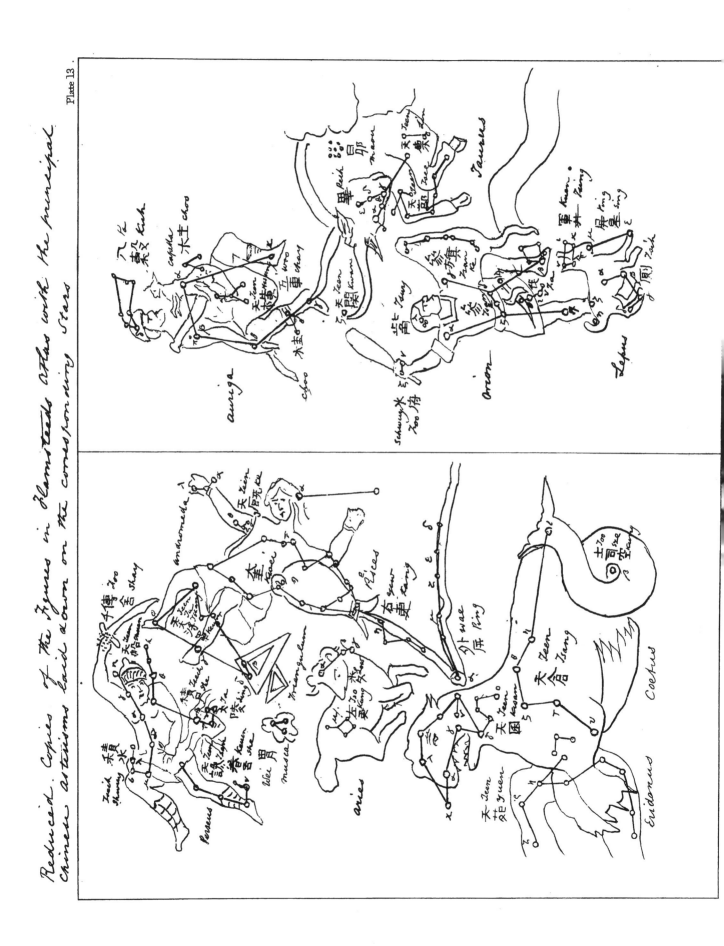

Plate 13

Reduced Copies of the figures in Flamsteed's Atlas with the principal Chinese Astronomical names laid down on the corresponding stars

Plate 14.

Plate 15.

Plate 16.

Printed in the United States
By Bookmasters